# The Pythagorean Theorem

ALFRED S. POSAMENTIER

# THE PYTHAGOREAN THEOREM

## The Story of Its *Power and Beauty*

AFTERWORD BY
DR. HERBERT A. HAUPTMAN, NOBEL LAUREATE

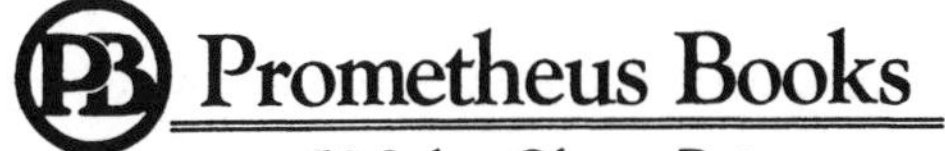

59 John Glenn Drive
Amherst, New York 14228–2119

Published 2010 by Prometheus Books



Inquiries should be addressed to
Prometheus Books
59 John Glenn Drive
Amherst, New York 14228-2119
VOICE: 716-691-0133
FAX: 716-691-0137
WWW.PROMETHEUSBOOKS.COM

14 13 12 11 10 5 4 3 2 1

Library of Congress Cataloguing-in-Publication Data

Posamentier, Alfred S.
The Pythagorean theorem : the story of its power and beauty
p. cm.
Includes bibliographical references and index.
ISBN 978-1-61614-181-3 (cloth : alk. paper)
1. Pythagorean theorem. I. Title.

QA460.P8 P67 2010
516.22–dc22 2010006838

Printed in the United States of America on acid-free paper

To Barbara, for her inspiration, support, and patience that made this book possible.

To my children and grandchildren, David, Lisa, Danny, Max, Sam, and Jack, whose futures are unbounded.

And in memory of my beloved parents, Alice and Ernest, who never lost faith in me.

Alfred S. Posamentier

# Contents

# Acknowledgments

To properly present the wide variety of Pythagorean appearances requires the expertise of specialists of these various fields. I was fortunate to obtain the support of such specialists for the following chapters: Professor Manfred Kronfellner of the Vienna University of Technology (Austria), a specialist in the history of mathematics, was instrumental in the preparation of chapter 1. Chapter 6 was contributed by an expert musicologist, Dr. Chadwick Jenkins, assistant professor of music at the City College of the City University of New York. The chapter on Pythagorean fractals (chapter 7) was written by two mathematicians at Central Michigan University: Dr. Ana Lúcia B. Dias and Dr. Lisa DeMeyer. Sincere gratitude is extended to these academics. As with any technical manuscript, careful proofreading with substantive comments and an eye toward clarity of presentation is paramount to the intelligibility of the work. I thank Peter Poole and Francisco Ruiz for their many helpful comments and I especially thank Professor (Emeritus) Michael Engber of the City College of the City University of New York, Professor Bernd Thaller of the Karl Franzens Universität Graz (Austria), and Professor Alfonso Bravo-Leon of the Universidad de Sevilla (Spain) for their meticulous proofreading of the entire book and the many insightful suggestions offered. Once again excellent editing by Peggy Deemer and Linda Regan is very much appreciated.

# Introduction

What did Pythagoras, Euclid, and US President James A. Garfield have in common? You were right if your answer was that they each proved the Pythagorean Theorem in a different way.

It is not uncommon at a social gathering to hear negative remarks about having to learn mathematics in school—especially when there is a mathematician in the group who represents the field. What is worse is when such well-educated folks exude pride in having been weak mathematics students in school. Some will claim to remember hardly anything from their school days' study of mathematics, yet will recall that "*a* squared plus *b* squared equals *c* squared" ($a^2 + b^2 = c^2$). This might be in part because this relationship used the first three letters of the alphabet. With some memory strain, some will be able to recall the author of this relationship: Pythagoras. They might then recall the name of this relationship: the Pythagorean Theorem. But what this relationship means is unfortunately not secure in the memory of most individuals. In fact, this famous theorem, which we clearly associate with geometry, is also the basis for the field of trigonometry and finds its way into countless other areas, such as art, music, architecture, and various fields of mathematics—principally the study of numbers. How did this relationship evolve? Why has this relationship fascinated innumerable generations? Who was this colorful and controversial man named Pythagoras? These are just a few of the tantalizing questions that we will consider in this expansive survey of this most popular mathematical relationship.

In its most basic sense, the Pythagorean Theorem states that if you draw a square on each side of a right triangle as in figure I-1, the sum of the areas of the two smaller squares (i.e., those on the perpendicular sides—called the legs of the right triangle) is equal to the area of the square drawn on the longest side, called the hypotenuse.

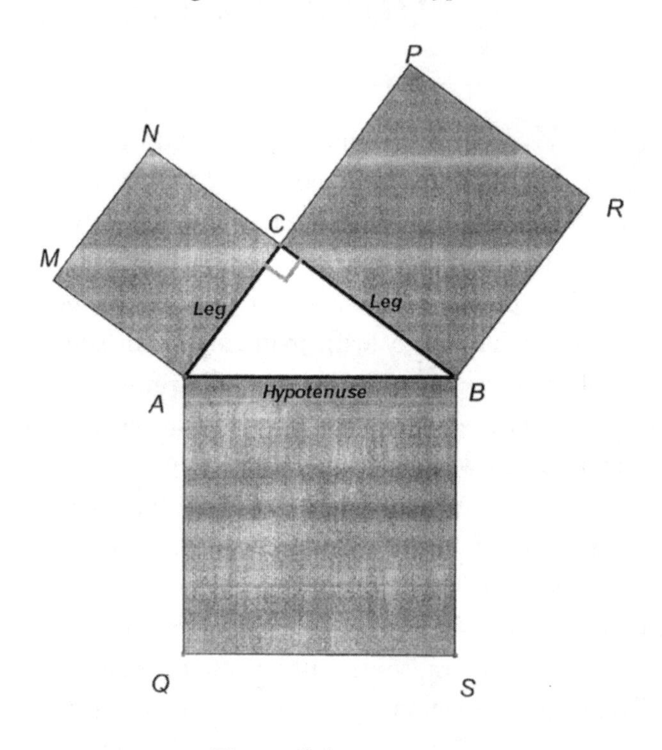


**Figure I-1**

While we can never be certain who first established this relationship among the sides of a right triangle, those in Western culture attribute this relationship to Pythagoras (ca. 575–495 BCE) and his school of followers, who imbued this phenomenal result with mystical meaning.

This relationship can be seen in many forms in our society. Take, for example, a tiled floor as shown in figure I-2. The number of congruent shaded and unshaded triangles in the squares on the two legs of the right triangle is equal to the number of the shaded and unshaded triangles in the square on the hypotenuse of the triangle.

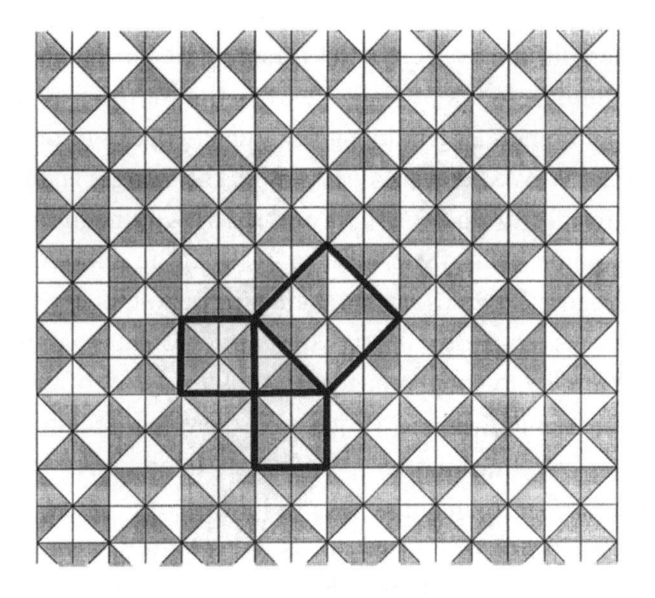

**Figure I-2**

What has dominated the study of the Pythagorean Theorem has been finding new proofs and applications of this relationship. Finding original proofs of the Pythagorean Theorem has fascinated mathematicians and amateur math enthusiasts for centuries. Currently more than five hundred proofs have been published establishing the truth of this most famous theorem. We will explore some of the more noteworthy proofs: those that are very succinct, those that are extremely clever, and those that demonstrate true elegance! We will also trace the omnipresence of this simple, yet powerful, theorem that has had a major impact on mathematics and so many other disciplines. Some of the other work attributed to Pythagoras—such as the field of music—will also be examined in the chapters that follow.

The shapes of right triangles are clearly dependent on the length of the sides. When the side lengths of two triangles are in proportion, they are similar. This is particularly true when the lengths of the sides of one triangle are a multiple of the lengths of the sides of the other triangle. If all sides are natural numbers,[1]

---

1. The *natural numbers* are the counting numbers 1, 2, 3, 4, ....

then the lengths of the sides of at least one of these triangles will have a common factor. However, of particular interest is when there is no common factor among the integer lengths of the sides of a right triangle. We call this a *primitive* right triangle and the three side lengths form what we call a primitive Pythagorean triple. There are some beautiful properties that can be discovered from these Pythagorean triples, as we shall explore in this book. For example, let us consider the most popular Pythagorean triple: (3, 4, 5). For this to be a Pythagorean triple, the sum of the squares of the first two numbers must equal the square of the third number, that is, $a^2 + b^2$ must equal $c^2$. That it does, since $3^2 + 4^2 = 9 + 16 = 25 = 5^2$. Exploring the product of the three numbers forming a Pythagorean triple is also interesting; in this case $3 \cdot 4 \cdot 5 = 60$. We will later show that the product of the members of any Pythagorean triple will always be a multiple of 60. We will also show you how to generate other Pythagorean triples and mine the riches they offer.

One of the interesting consequences of the Pythagorean Theorem is its connection to the discovery of the existence of irrational numbers, which are numbers that cannot be written in the form $\frac{a}{b}$, where $b$ does not equal 0. Some people think that there are no numbers other than whole numbers and fractions. In fact, there are infinitely many numbers that don't fit these forms. Some are called irrational numbers and others are called imaginary numbers. There are some geometric lengths that are not exactly measurable with a standard inch (or millimeter) ruler. However, they can be constructed accurately (such as a line segment whose length is $\sqrt{2}$ inches long) with the standard geometric construction tools: an unmarked straightedge and a pair of compasses. Such numbers were controversial at one time, but they gave rise to a whole new set of numbers that we now take for granted. These numbers cannot be represented as the *ratio* of two integers and therefore are not *rational*, and so are called *irrational.* Indeed, the fact that these numbers are called "irrational" is an indication of the brouhaha that accompanied their discovery. Numbers that could be written as fractions were called *rational* (i.e., written as a *ratio* of two numbers), and numbers that could not were considered to be *irrational.* As we use the Pythagorean Theorem, we will be employing both rational and irrational numbers.

The Pythagorean Theorem is also the key to analytic geometry, a study of geometry on a coordinate plane, such as is seen on graph paper. When we measure distances on this coordinate plane—also called the Cartesian plane after the French mathematician René Descartes, who introduced this approach to geometry—we need to use the Pythagorean Theorem. Additionally, the entire field of trigonometry relies on the truth of the Pythagorean Theorem. These are just some appearances of this famous theorem. We will discover a treasure trove of others together.

In this book, we will examine the wide variety of applications of the Pythagorean Theorem, many of which were unanticipated at its inception and are now deeply ingrained in our mathematics. We will also delve into other nonmathematical fields to observe the unexpected occurrences of the Pythagorean Theorem, such as how it plays a part in our popular culture (e.g., the scarecrow in *The Wizard of Oz* spouts the Pythagorean Theorem when he realizes that he's had a brain all along). Furthermore, in the fields of music and fractal art, we will also observe the attention given to Pythagoras's work.

Join us as we visit and explore the power and glory of the most famous theorem in geometry and possibly in all of mathematics. We will take you on a fascinating journey through some truly amazing mathematics.

# Chapter 1

# Pythagoras and His Famous Theorem

As we embark on our exploration of the Pythagorean Theorem, we are faced with some questions. Chief among them is why is the relationship that historically bears that name—the Pythagorean Theorem—so important? There are many reasons: perhaps because it is easy to remember; perhaps because it can be easily visualized; perhaps because it has fascinating applications in many fields of mathematics; or perhaps because it is the basis for much of mathematics that has been studied over the past millennia. We shall explore these aspects in the chapters that follow. But, perhaps it is best to begin at its roots, with the mathematician whom we credit as being the first to prove this theorem, and examine the man himself, his life, and his society.

The first biography of Pythagoras was written about eight hundred years after his death by Iamblichus, one of many Pythagoras enthusiasts, who tried to glorify him. And, although Pythagoras has been mentioned numerous other times throughout history, by well-known writers such as Plato, Aristotle, Eudoxus, Herodotus, Empedocles, and others, we still do not have reliable information about him. Some of his contemporary followers actually believed that he was a demigod, a son of Apollo—a conviction they supported by noting that his mother was said to be a very beautiful woman. Some reported that he even worked wonders.

But even though he was called the greatest mathematician and philosopher of antiquity by some, he was not without critics who tried to revile him. They say that he was merely the founder and chief of a sect—the Pythagoreans—and that the many scientific results that came from them were written by the members of the sect and dedicated to its leader, and thus were not the work of Pythagoras himself. The critics considered him a collector of facts without any deeper understanding of the related concepts, and therefore felt he did not really contribute to a deep understanding of mathematics. Similar criticism also was aimed at such luminaries as Plato, Aristotle, and Euclid. We must remain mindful of these uncertainties when we consider the "facts" about Pythagoras's life and work.

Pythagoras was born circa 575 BCE[1] on the island Samos (located off the west coast of Asia Minor). His initial and perhaps most influential teacher was Pherecydes, who was primarily a theologist who taught him religion and mysticism along with mathematics. As a young man he traveled to Phoenicia, Egypt, and Mesopotamia, where he advanced his knowledge of mathematics and also pursued a variety of other interests, such as philosophy, religion, and mysticism. Some biographers believe that, while in his late teens, Pythagoras traveled first to Miletus, a coastal town in Asia Minor near Samos, where he continued his studies in mathematics under the tutelage of the famous philosopher and mathematician Thales of Miletus. It is very likely that he also attended lectures from another Miletic philosopher, Anaximander, who further inspired Pythagoras in geometry. When he returned to Samos, the tyrant Polycrates,[2] who ruled Samos from 538 to 522 BCE, had come to power. It is not clear if Pythagoras disagreed with Polycrates' leadership. But soon after returning home, he

---

1. Many history books offer various dates for Pythagoras's birth, between 600 and 570 BCE.

2. Polycrates was made famous in modern times by the German writer Friedrich Schiller (1759–1805), who, in 1797, wrote the ballad "The Ring of Polycrates."

moved to Croton (today, Crotone in southern Italy), about 530 BCE, a region that had had a considerable Greek population since the eighth century. There he founded a community—or society—whose main interests were religion, mathematics, astronomy, and music (acoustics). The Pythagoreans' conviction that all aspects of nature and the universe could be explained and expressed by means of the natural numbers and the ratios of numbers suffered a setback when they learned that the emblem of their community, the pentagram, contradicted their core numerical principles.[3]

The Pythagoreans tried to explain the nature of the world and the universe with the help of numbers. In particular, they studied vibrating strings and found that two strings sound harmonious if their lengths can be expressed as the ratio of two small natural numbers such as 1:2, 2:3, 3:4, 4:5, and so on. They came to believe that the entire universe is ordered by such simple relations of natural numbers. This ties in with their study of the three most popular means: the arithmetic mean, the geometric mean, and the harmonic mean, which relate to each other. We will visit these means in chapter 5.

One of the core beliefs of the Pythagoreans is that there is a strong connection between religion and mathematics. They believed that the sun, the moon, the planets, and the stars were of a divine nature and therefore they could move only along circular paths. Furthermore, they believed that the movements of these bodies caused sounds of different frequencies because of their different velocities, which in turn depended on their radii. These sounds were said to generate a harmonic scale, which they called the "harmony of the spheres." Yet they be-

---

3. One consequence of their conviction that all aspects of nature can be described by means of numbers would have been that, in particular, every two lines have a common measure: that is, that they are commensurable. Two magnitudes $a$, $b$ are called commensurable if there exists a magnitude $m$ and whole numbers $\alpha$, $\beta$ so that $a = \alpha \cdot m$ and $b = \beta \cdot m$. But in a pentagram (i.e., a regular five-corner star shape) the sides and the diagonals are *not* commensurable! It is said that Hippasus of Metapontum, a student of Pythagoras, discovered this fact and mentioned it to people outside the community. This was regarded as a violation of the pledge of secrecy, so Hippasus was subsequently banned from the community. Some say that he died in a shipwreck, which was regarded as a punishment from the gods for his sacrilege. Another version tells that he was killed by other members of the society.

lieved that man cannot actually hear this sound, as it surrounds humans constantly from birth. Even the great scientist Johannes Kepler (1571–1630) was sometimes characterized as a late Pythagorean since he believed that the diameters of the orbits of the planets could be explained by inscribed and circumscribed Platonic solids[4] (see figure 1-1), an idea he published in his work *De Harmonice Mundi* (*About the Harmony of the World*).

By investigating the courses of celestial bodies the Pythagoreans sought to purify their souls and to prepare them for their final passage into the heavens. Before that final stage—so they believed—their souls would transmigrate, not only from human to human but also into animals. Therefore, among their other rules, those of modesty, discipline, and secrecy, they were said to have enforced a ban on sacrificing animals and eating meat, for they thought that the soul of a deceased person might be in the animal. To further their ability to focus on their beliefs, they also refrained from eating beans, as that produced flatulence and interfered with intellectual thinking. Some biographers contend, however, that the Pythagoreans enforced their animal-sacrificing ban only on some animals—namely, those that they believed had a soul. In particular, one anecdote reports that the Pythagoreans sacrificed twenty oxen whenever they came up with and proved a mathematical concept.

In contrast to the Pythagoreans, the philosophers Anaxagoras and Democritus believed that planets and stars were only glowing stones. Anaxagoras was sentenced to death because of his so-called godlessness in espousing this belief; but his sentence was commuted to banishment after the intervention of the revered statesman Pericles.

---

4. *Platonic solids* are solids whose surfaces consist of regular polygons of the same type. There exist only five Platonic solids: the tetrahedron (a pyramid made of equilateral triangles), the cube (made of six squares), the octahedron (a double pyramid made of eight equilateral triangles), the dodecahedron (made of twelve pentagons), and the icosahedron (made of twenty equilateral triangles).

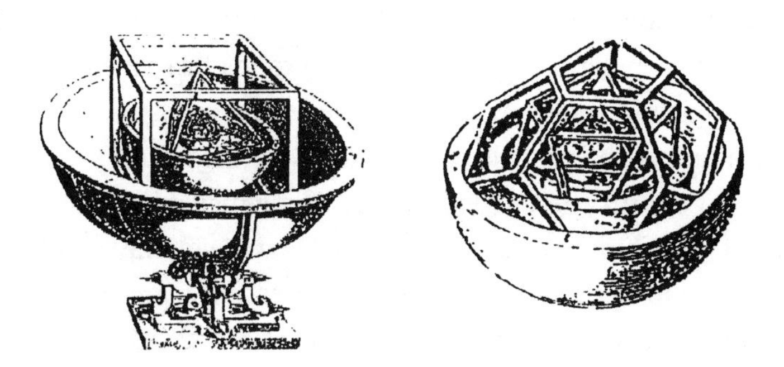

**Figure 1-1**

Part of the reason Pythagoras gained such a large following was because he was an eloquent speaker; in fact, four of his speeches, given to the public in Croton, are still remembered today. The Pythagoreans also gained political influence in that region, even over the non-Greek population. But sometimes—as is frequent in politics—they faced resistance and animosity. Later (in approximately 510 BCE), when the Pythagoreans got involved in various political disputes, they were expelled from Croton. The displaced Pythagoreans tried to move to other towns, such as Locri, Caulonia, and Tarentum, but the people in these towns did not allow them to settle. Finally, in Metapontum, they found their new home. Here Pythagoras died of old age around 495 BCE.

As there was no appropriately charismatic leader to succeed Pythagoras, the Pythagoreans split up into several small groups and tried to continue their tradition, while continuing to exert political influence in various towns in southern Italy. They were rather conservative and well connected to established influential families, which got them into conflict with their common counterparts. As soon as their opponents gained the upper hand, bloody persecutions of the Pythagoreans began. Given the political situation, many of them emigrated to Greece. This was—more or less—the end of the Pythagoreans in southern Italy. Very few individuals tried to continue the tradition and to advance the Pythagorean ideals. Two groups that persisted were the Acusmatics and the Mathematics. The former believed in acusma (i.e., what they had *heard* Pythagoras say) and did not give any further explanations. Their only justification was "He said it." This gave Pythagoras a level of importance, or popularity, in his day, which to another extent still exists today. In contrast to the Acusmatics, the Mathematics tried to develop his ideas further and provide precise proofs for them.

One of the very few Pythagoreans who remained in Italy was Archytas of Tarentum (ca. 428–350 BCE). He was not only a mathematician and philosopher but also a very successful engineer, statesman, and military leader. He befriended Plato in about 388 BCE, giving rise to the belief that Plato learned the Pythagorean philosophy from Archytas, and that is why he discussed it in his works. Aristotle, who was first a student in Plato's academy but soon became a teacher there, wrote rather critically about the Pythagoreans. While Plato may have adopted many ideas from the Pythagoreans, such as the divine nature of planets and stars, in other cases he disagreed with them. Plato mentioned Pythagoras only once in his books, but not as a mathematician, despite his being in close contact with all of the mathematicians of his time and holding them in high regard. It is probable that Plato did not consider Pythagoras a proper mathematician. Similarly, Aristotle also mentioned the Pythagoreans, but said almost nothing about Pythagoras.

In the fourth century BCE the Greeks distinguished between "Pythagoreans" and "Pythagorists." The latter were extremists of the Pythagorean philosophy and consequently often the target of sarcasm because of their unusual ascetic lifestyle. Still, among the Pythagoreans there were some members who were able to command respect from outsiders.

After the fourth century BCE, the Pythagorean philosophy disappeared from sight until the first century CE when Pythagoras came into vogue in Rome. This "Neo-Pythagoreanism" remained alive in subsequent centuries. In the second century CE Nicomachus of Gerasa wrote a book about the Pythagorean number theory whose Latin translation by Boethius (ca. 500 CE) was widely distributed. Today, Pythagorean ideas permeate our thinking in a variety of fields, as we will see.

## The Pythagorean Theorem

Let us now focus on the geometric relationship that made Pythagoras famous in today's world and that, of course, bears his name. We would do well to consider his prominent role (or that of his society) in the development of this amazing relationship.

Although the relationship was already known before Pythagoras (as you will see in the ensuing pages), it is appropriate that the theorem should be named for him, since Pythagoras (or one of the Pythagoreans) was the first to give a proof of the theorem—at least as far as we know. Historians suppose that he used the squares as shown in figures 1-2 and 1-3—perhaps inspired by the pattern of floor tiles. We will briefly demonstrate the proof here.[5]

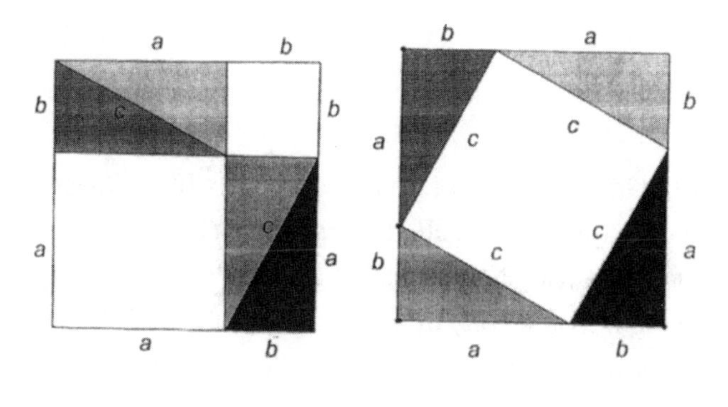


Figure 1-2

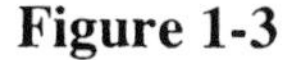

Figure 1-3

To show that $a^2 + b^2 = c^2$, you need only subtract the four right triangles, with sides $a$, $b$, and $c$ from each of the two larger squares, so that in figure 1-2 you end up with $a^2 + b^2$, and in figure 1-3 you end up with $c^2$. Therefore, since the two original squares were the same size and we subtracted equal quantities from each, we can conclude that $a^2 + b^2 = c^2$, which is shown in figure 1-4 with the two figures of the same area.

---

5. The proof will be shown in greater detail in chapter 2 as demonstration 1 (page 38).

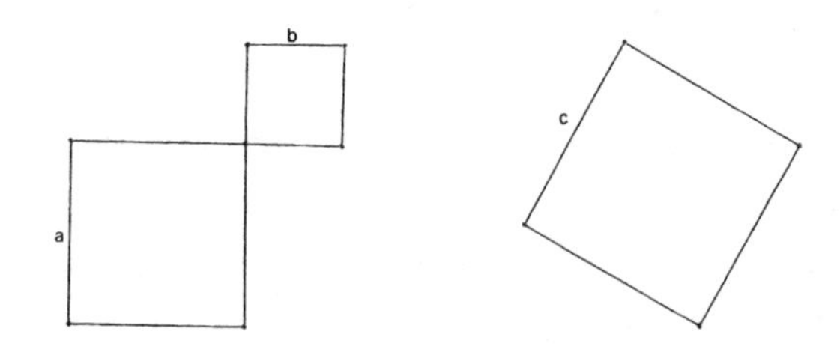


**Figure 1-4**

To prove a theorem is one thing, but to come up with the idea establishing this geometric relationship is quite another. It is likely that Pythagoras learned about this relationship on his study trip to Egypt and Mesopotamia, where this concept was known and used in construction for special cases.

## Egypt

During his travels to Egypt, Pythagoras probably witnessed the measuring method of the so-called Harpedonapts (rope stretchers). They used ropes tied with 12 equidistant knots to create a triangle with one side of length 3 units, one of 4, and a third side of 5, knowing that this enabled them to "construct" a right angle. (See figure 1-5.)

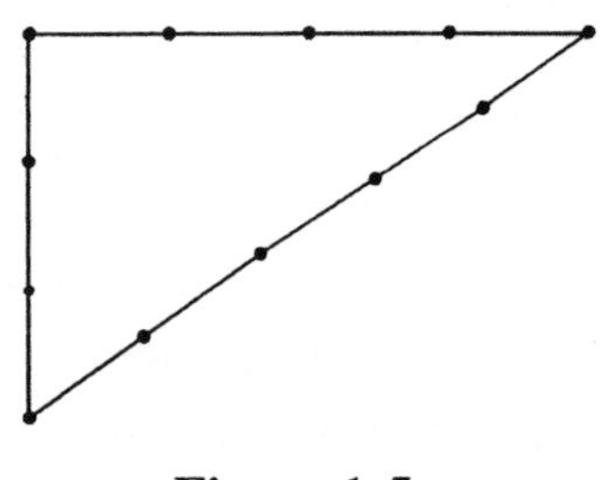

**Figure 1-5**

They applied this knowledge to survey the banks of the river Nile after the annual floods in order to rebuild rectangular fields for the farmers. They also employed this method in laying the foundation stones of temples. To the best of our knowledge, the Egyptians did not know of the generalized relationship given to us by the Pythagorean Theorem. They seem to have only known about the special case of the triangle with side lengths 3, 4, and 5, which produced a right triangle. This was arrived at by experience and not by some sort of formal proof.

## Mesopotamia

In Mesopotamia, mathematicians were even able to produce further triples of numbers, fulfilling the Pythagorean condition of $a^2 + b^2 = c^2$, as we can see on a Babylonian clay tablet from ca. 1800 BCE, known as the Plimpton 322.[6] (See figure 1-6.) The tablet was part of a collection of about a half million of such tablets found in the mid-nineteenth century as a result of Mesopotamian digs, of which about three hundred were identified as having mathematical significance. The tablet is written in Old Babylonian (or cuneiform) script and uses the sexagesimal system (base 60). It shows us the high level of mathematical knowledge that existed well before the Greeks.

**Figure 1-6**

6. This tablet is in the permanent collection of the Columbia University Library, New York.

The two shaded columns of figure 1-7 translate the Babylonian numerals to our base-10 system and give a strong indication of their knowledge of the Pythagorean triples.[7] These two columns list the leg and hypotenuse of several Pythagorean triples.

Here, we notice that the three left-hand numbers in each row satisfy the Pythagorean Theorem, $a^2 + b^2 = c^2$, and are called Pythagorean triples.

| $a$ | $b$ | $c$ | $m$ | $n$ |
|---|---|---|---|---|
| 120 | 119 | 169 | 12 | 5 |
| 3456 | 3367 | 4825 | 64 | 27 |
| 4800 | 4601 | 6649 | 75 | 32 |
| 13500 | 12709 | 18541 | 125 | 54 |
| 72 | 65 | 97 | 9 | 4 |
| 360 | 319 | 481 | 20 | 9 |
| 2700 | 2291 | 3541 | 54 | 25 |
| 960 | 799 | 1249 | 32 | 15 |
| 600 | 481 | 769 | 25 | 12 |
| 6480 | 4961 | 8161 | 81 | 40 |
| 60 | 45 | 75 | 2 | 1 |
| 2400 | 1679 | 229 | 48 | 25 |
| 240 | 161 | 289 | 15 | 8 |
| 2700 | 1771 | 3229 | 50 | 27 |
| 90 | 56 | 106 | 9 | 5 |

**Figure 1-7**

Pythagorean triples have also been discovered in northern Europe in megalithic rings, where they are displayed as triples of numbers that are, in large measure, accurate Pythagorean triples. However, in Babylonia we not only find Pythagorean triples but we also find problems, which can only be solved with a proper knowledge of the Pythagorean Theorem.

The Babylonians derived the *Pole against the wall problem.* If a pole of length 0;30 units slips down 0;6 units along the 0;30-unit

---

7. Four of the entries had errors, which we corrected in the chart. For each of the correct entries that follow, the error is given in parentheses: 4825 (11521), 481 (541), 161 (25921), and 106 (53).

wall, how far is the base of the pole from the base of the wall?[8] (See figure 1-8.)

You *make* 0;30 *hold itself*, you see 0;15.
You *tear out* 0;6 from 0;30, you see 0;24.
You *make* 0;24 *hold itself*, you see 0;9,36.
You *tear out* 0;9,36 from 0;15, you see 0;5,24.
0;5,24 has what *square side*? It *has* 0;18 *as square side*.
It has moved away 0;18 (*nindānu*) on the ground.

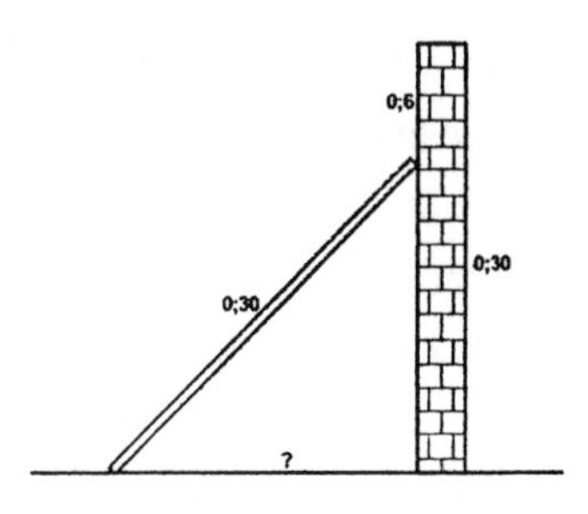


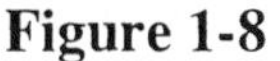
**Figure 1-8**

This calculation is seen geometrically in figure 1-9.

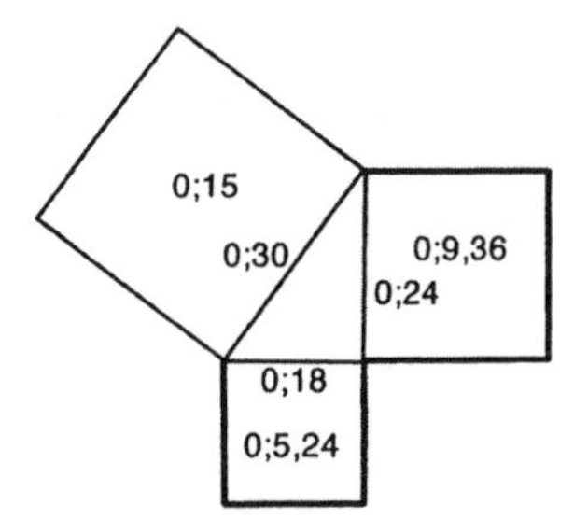


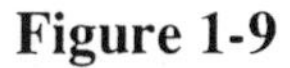
**Figure 1-9**

The calculation in our words and our number symbols would translate to the following:

Square 0.5, you will get 0.25.
Subtract 0.1 from 0.5, you will get 0.4.
Square 0.4, you will get 0.16.

---

8. The Babylonians used a number system with the base 60 (instead of 10 as we do). In particular, 0;30 means $0+\frac{30}{60}=0.5$, 0;6 means $0+\frac{6}{60}=0.1$, and 0;9,36 means $0+\frac{9}{60}+\frac{36}{60^2}$, etc.

Subtract 0.16 from 0.25, you will get 0.09.
0.09 is the area of a square, so its side is 0.3.

The stick has slipped by 0.3, which, when converted to the sexagesimal system, is 0;18.[9]

On another clay tablet, YBC 7289 (Yale University Babylonian Collection), we can see that the Babylonians already applied the Pythagorean Theorem to calculate a rather accurate approximation of $\sqrt{2}$. In figure 1-10, there are three pictures of this tablet; the first shows the original tablet, the second displays it with accentuated marking, and the third shows the values of the lengths of its sides. On this tablet there is a square, whose sides have length 1, with a value written along the diagonal. That is, there is an isosceles right triangle—half the square—where, by using the Pythagorean Theorem, we can determine that in that isosceles right triangle with legs of length 1, the hypotenuse is length $\sqrt{2}$, since $1^2 + 1^2 = 2 = c^2$, so that $c = \sqrt{2}$.

The number along the diagonal is: 1;24,51,10, which means:

$$1 + \frac{24}{60} + \frac{51}{60^2} + \frac{10}{60^3} = 1.414212963$$

This is a very close approximation of the value of $\sqrt{2} =$ 1.4142135… .

The second line, 42;25,35, is the product of this number, $\sqrt{2}$, with the length of the (upper-left) leg given as 30. That is, the second line's value is base 60, which is:

$$42 + \frac{25}{60} + \frac{35}{60^2} = 42.42638889, \text{ while } 30 \cdot \sqrt{2} = 42.42640687...$$

a very close approximation!

9. $0.3 = \frac{18}{60}$, which we write as 0;18.

**Figure 1-10**

## India

We assume—because of geographical reasons—that Pythagoras learned about the relationship in right triangles in Egypt or Mesopotamia. Some historians, however, argue that he may also have learned about it in India. Similarities were discovered between Indian philosophies and Pythagoras's principles. In the *Sulva Sutra* ("rules of the rope") of Baudhayana (about 800 BCE), we can already find problems that refer to the Pythagorean relation:

> A rope stretched along the length of the diagonal produces an area which the vertical and the horizontal sides make together.[10]

---

10. This means that if the side of a square is the length of the diagonal of a rectangle, then it has an area that equals the sum of the areas of the squares whose side-lengths are the sides of the rectangle.

And for the special case of an isosceles right triangle there is:

> A chord which is stretched across the square produces an area double the size of the original square.[11]

As a special case of the previous relationship, which was applied to an isosceles right triangle, we can observe that the folding and unfolding (shown in figure 1-11) is analogous to the argument that Socrates uses in his style of "teaching" Meno's slave in Plato's *Meno* (380 BCE), by taking a concrete example in a very simple situation.

These statements are the earliest recorded of the Pythagorean Theorem in Indian mathematical literature. Other books written in subsequent centuries contain similar statements; therefore, we can assume that Baudhayana's *Sulva Sutra* was well known in this region. One could further assume that Pythagoras may have become familiar with this idea when he came in contact with this culture.

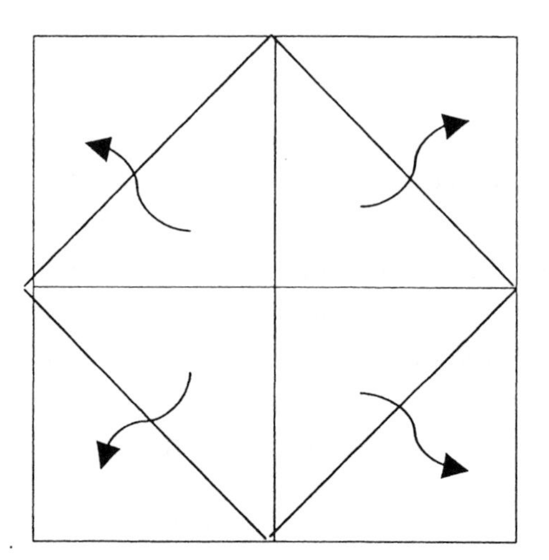

**Figure 1-11**

---

11. This recognizes that the diagonal of a unit square must be equal to $\sqrt{2}$, since that length—as the side of a square—would produce an area of 2, or twice that of the original square.

## China

Despite a historical emphasis on Western culture, one must also consider what happened in the Far East. Was Pythagoras the first to discover the theorem that bears his name or was he anticipated in the Far East? In the book *Zhou Pei Suan Jing*, one of the oldest Chinese books on mathematics (perhaps even the oldest), we can also find the Pythagorean Theorem. It is not clear when it was written. Some historians believe it was in the twelfth century BCE, while others place the work in the first century BCE. So it is not clear whether the Indian mathematicians learned the Pythagorean Theorem from the Chinese (or vice versa) or whether they discovered the theorem independently. The last assumption seems most likely, because in Chinese mathematics, the Pythagorean Theorem was more of an exercise in arithmetic, whereas in Indian or Babylonian mathematics, the principles of the Pythagorean Theorem arose from geometrical measurement.

Almost as old as the *Zhou Pei Suan Jing* is a book called *Jiuzhang Suanshu* (*Nine Chapters on the Mathematical Arts*), which is probably the most influential of all ancient Chinese books on mathematics. It contains a collection of 246 problems on surveying, engineering, and other subjects that relate to mathematics.

The Pythagorean Theorem is called *Gougu*. *Gou* means the shorter leg of a right triangle (originally: of a carpenter's square) and *gu* refers to the longer leg. The hypotenuse is called *xien*, which translates to "spanned chord."

One of the problems is that of the Broken Bamboo. In figure 1-12 we see that a bamboo pole is 10 feet tall. The upper end is broken and the top reaches the ground 3 feet from the stem. The question asked is to find the height of the break. Such is the ubiquity of this famous theorem.

In contrast to other cultures that did not provide justifications for the Pythagorean Theorem, the Chinese gave a "proof," at least for the special case of the 3-4-5 triangle.

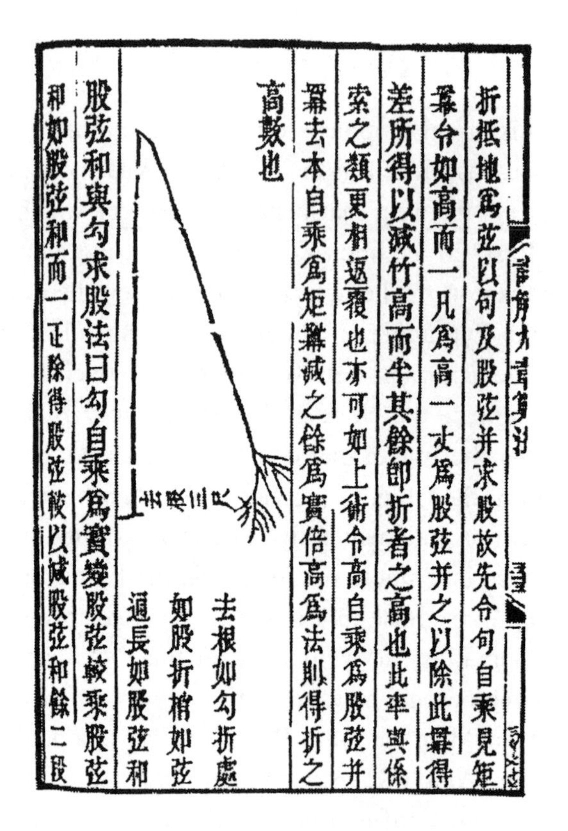

折抵地爲弦以句及股弦并求股故先令句自乘見矩
冪令如高而一凡爲高一丈爲股弦并之以除此冪得
差所得以減竹高而半其餘卽折者之高也此率與係
索之類更相返覆也亦可如上術令高自乘爲股弦并
冪去本自乘爲矩冪減之餘爲實倍高爲法則得折之
高數也



股弦和與勾求股法曰勾自乘爲實變股弦較乘股弦
和如股弦和而一正除得股弦較以減股弦和餘二段

**Figure 1-12**

If you subtract the four dark right triangles in both squares of figure 1-13, then you will find you are left with $c^2$ in the first square and $a^2 + b^2$ in the second square.

Regardless of where it originated, the Pythagorean Theorem has fascinated mathematicians and amateurs alike through the ages. As you will see in chapter 2, such famous people as Plato, Euclid, Aristotle, Leonardo da Vinci, and US president James A. Garfield produced original proofs of this time-honored theorem. Mathematics enthusiasts have searched for proofs of this famous theorem for millennia. In fact, in 1940, Elisha Scott Loomis (1852–1940), then professor emeritus of mathematics at Baldwin-Wallace College, published the second edition of his *Pythagorean Proposition*, which

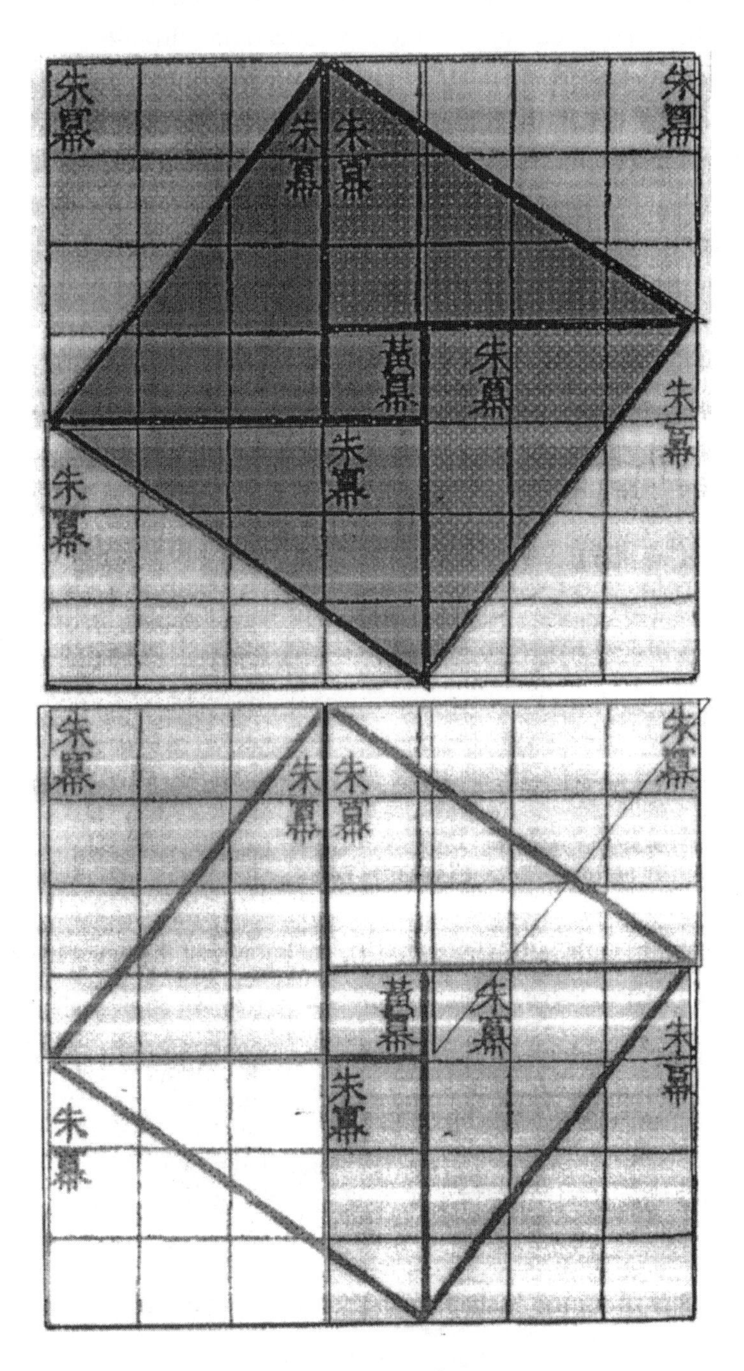


Figure 1-13

contains 367 different proofs of the Pythagorean Theorem. Since then there have been many others produced and proudly published in mathematics journals. Yet, as Loomis clearly noted, not one of the proofs of the Pythagorean Theorem that he presented in his book used trigonometry since, as he correctly pointed out, the field of trigonometry depends on the Pythagorean Theorem for its primary relationship, namely, that $sin^2 A + cos^2 A = 1$. Therefore, using trigonometry to prove the Pythagorean Theorem would be tantamount to circular reasoning, since proving a theorem with a "tool" that depends on the theorem is not proper logic. The establishment of $sin^2 A + cos^2 A = 1$ can be easily shown by applying the Pythagorean Theorem to the sine and cosine ratios for right triangle *ABC* shown in figure 1-14.

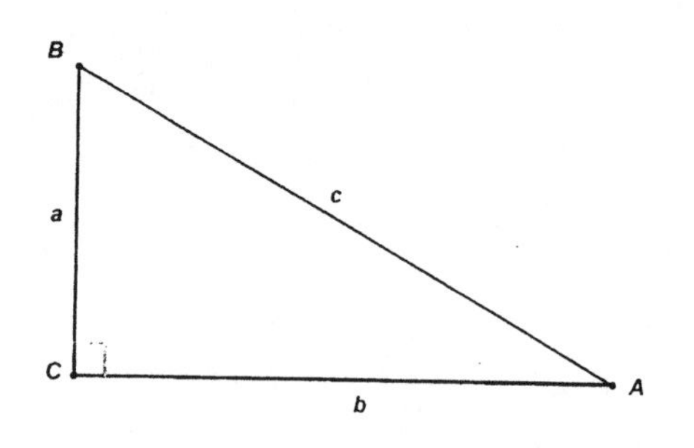


**Figure 1-14**

For right triangle *ABC* (figure 1-14), we know that the sine ratio is $sin\, A = \frac{a}{c}$, and the cosine ratio is $cos\, A = \frac{b}{c}$. As we said, the basic trigonometric relationship that underpins the field is $sin^2 A + cos^2 A = 1$. From the above defining ratios we get

$$sin^2 A + cos^2 A = \left(\frac{a}{c}\right)^2 + \left(\frac{b}{c}\right)^2 = \frac{a^2}{c^2} + \frac{b^2}{c^2} = \frac{a^2+b^2}{c^2}$$

However, from the Pythagorean Theorem we have $a^2 + b^2 = c^2$. Therefore, $\frac{a^2+b^2}{c^2} = 1$, and then $sin^2 A + cos^2 A = 1$.

To further appreciate this revelation, consider the words of the great scientist Albert Einstein:

> I remember that an uncle told me the Pythagorean Theorem before the holy geometry booklet had come into my hands. After much effort I succeeded in "proving" this theorem on the basis of the similarity of triangles. . . . For anyone who experiences [these feelings] for the first time, it is marvelous enough that man is capable at all to reach such a degree of certainty and purity in pure thinking as the Greeks showed us for the first time to be possible in geometry.[12]

In this spirit of discovery, we now embark on a journey guided by the Pythagorean Theorem and the work of Pythagoras as they apply to various aspects of our modern body of knowledge in mathematics and beyond.

---

12. From pp. 9–11 in the opening autobiographical sketch of *Albert Einstein: Philosopher-Scientist*, ed. Paul Arthur Schilpp (New York: Tudor, 1951).

# Chapter 2

# Proving the Pythagorean Theorem without (Many) Words

Our fascination with the Pythagorean Theorem is that it is one of the fundamental building blocks of mathematics. Clearly geometry, as we know it today, could not exist without this significant theorem. Trigonometry, for example, is essentially based on this important theorem. Further fueling our fascination with the Pythagorean Theorem is the constant motivation that both mathematicians and amateurs have in finding new proofs to establish the truth of it. As we noted, there are geometric proofs—some quite complex—and there are algebraic proofs, but there are no proofs using trigonometry since this field is based largely on the Pythagorean Theorem and would result in circular reasoning.

From time to time—to the present day—there arise what are believed to be new proofs or demonstrations of the Pythagorean Theorem published in professional journals. A major contribution to the collection of Pythagorean Theorem proofs was contributed by Elisha Scott Loomis (1852–1840), whom we discussed in the last chapter. Loomis wrote a book, *The Pythagorean Proposition*,[1] with 367 proofs of this famous theorem. His methods were all algebraic and geometric.

The most delightful proofs of the Pythagorean Theorem can be shown with almost no explanation other than symbols. They speak for themselves. These proofs often demonstrate unexpected ingenuity and give a deeper insight into the theorem, contributing significantly to our appreciation of it. As you read some unusually clever proofs of the theorem, some comments will be offered to assist you to explain what is being done in these largely visual demonstrations. In this chapter, we will also provide some rather ingenious proofs (or demonstrations) that sometimes use highly imaginative diagrams that lead smoothly to the desired result. As you consider these demonstrations, especially the visual ones, follow the discussion provided since it will give you a nice window to the beauty of geometry. We discussed the method often thought to be the one that Pythagoras actually used, on page 23.

## Demonstration 1

A visual proof[2] is attributed to the Hindu mathematician Bhaskara (ca. 1150). He began with a right triangle with sides of length $a$, $b$, and $c$ (figure 2-1).

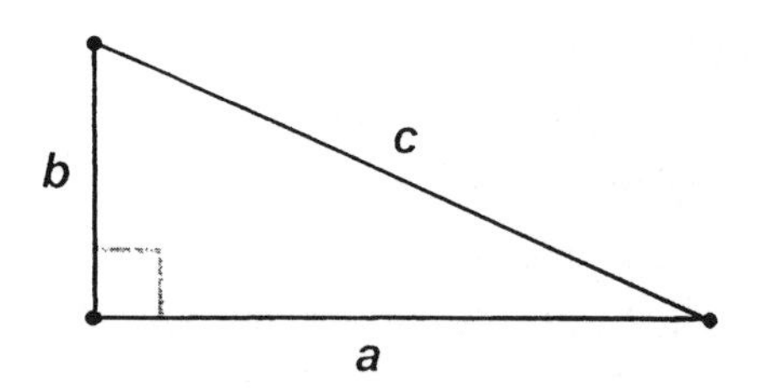


**Figure 2-1**

1. Originally published in 1940, it was republished by the National Council of Teachers of Mathematics (Reston, VA) in 1968.

2. Howard Eves, *Great Moments in Mathematics (Before 1650)* (Washington, DC: Mathematical Association of America, 1980), pp. 29–31.

Bhaskara merely drew the square with a side length of the hypotenuse of the given right triangle (figure 2-1) and partitioned it into four congruent right triangles (also congruent to the original right triangle) and a smaller square (see figure 2-2). He then repositioned these five parts as in figure 2-3. With the word "behold" beside it, he considered the demonstration of the Pythagorean Theorem complete. The explanation of this is that figures 2-2 and 2-3 have equal areas. In figure 2-2, the area of the square with side length $c$ is simply $c^2$. The area shown in figure 2-3 is composed of two squares—one with sides of length $a$ and one with sides of length $b$. The sum of the areas of these two squares (which is the entire figure) is $a^2 + b^2$. Therefore, since the areas shown in figures 2-2 and 2-3 are equal—each is composed of the same four congruent right triangles and square—we can conclude that $a^2 + b^2 = c^2$.

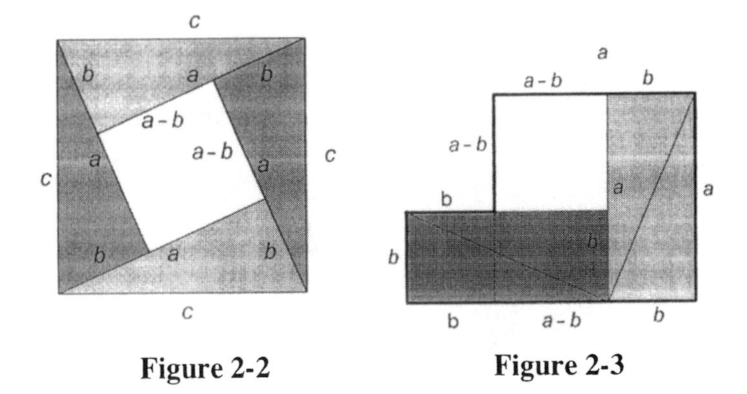


Figure 2-2 Figure 2-3

Before we embark on the next demonstration of the Pythagorean Theorem, we ought to consider a simple relationship about area comparisons. In figure 2-4a, we notice that the square *ACNM* and the parallelogram[3] *ACEH* have the same base, *AC*, and the same altitude, *NC* (since the lines *MK* and *AP* are parallel). Therefore, their areas are equal (remember the formula for the area of a square and a parallelogram is $A = bh$).

---

3. A *parallelogram* is a quadrilateral with both pairs of opposite sides parallel.

Now consider the triangle $ACE$ in figure 2-4b. Recall that the formula for the area of a triangle is $A = \frac{1}{2}bh$. The square $ACNM$ and the triangle $ACE$ have the same base and altitude, so the area of triangle $ACE$ is half that of square $ACNM$. We will be using these relationships in some of the following demonstrations.

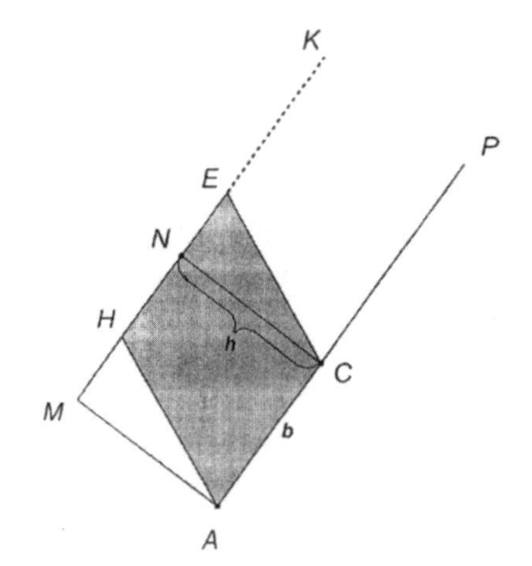


**Figure 2-4a**

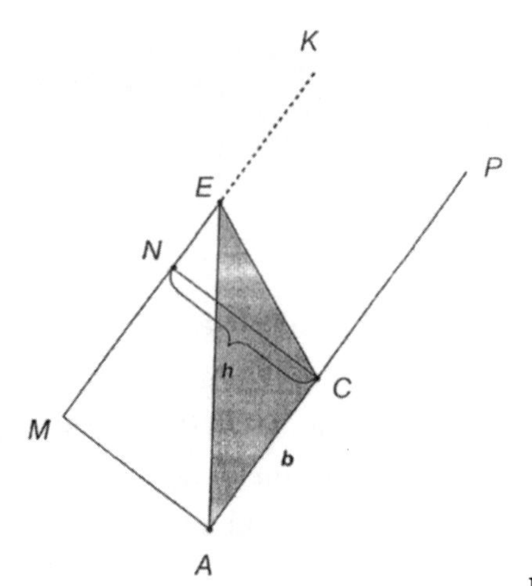


**Figure 2-4b**

## Demonstration 2

We can also prove the Pythagorean Theorem by moving parts of a figure around and show that the sum of the areas of the squares on the legs of a right triangle is the same as the area of the square on the hypotenuse. In figure 2-5, we begin with the three squares drawn on the sides of right $\Delta ABC$.[4] Our approach here will be to slide the shaded regions as shown in figures 2-5–2-9. We will provide the explanations for each move.

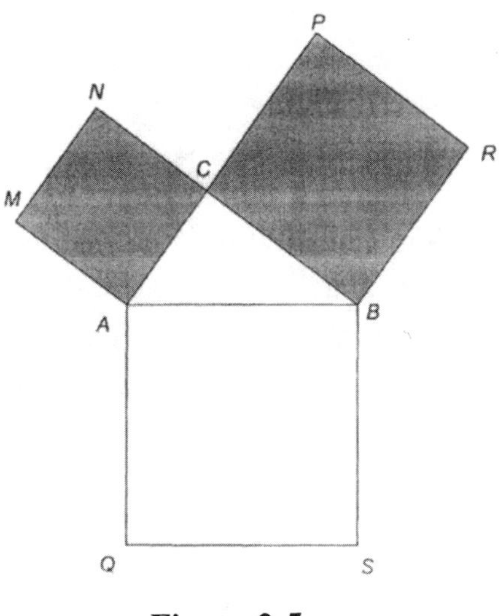


**Figure 2-5**

First we extend *NM* and *RP* to meet at point *K* as shown in figure 2-6. We notice that the two shaded parallelograms have the same areas as the squares with which they share their base. That is, the area of parallelogram *ACEH* is equal to the area of square *ACNM*—since they share base *AC* and altitude *CN*.

For the same reason, the area of parallelogram *BCFG* is equal to the area of square *BCPR*.

4. Eves, *Great Moments in Mathematics (Before 1650)*, pp. 31, 33.

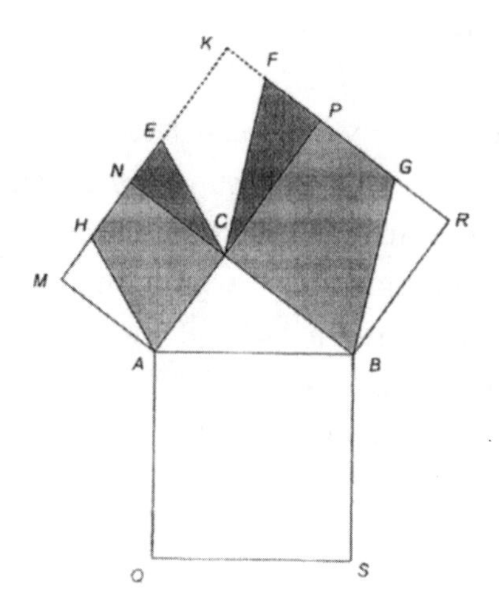


**Figure 2-6**

Sliding the parallelograms further along until they touch as shown in figure 2-7, we can still conclude that the two parallelograms have the same areas as the squares, respectively. Therefore, the sum of the areas of the two shaded parallelograms is equal to the sum of the areas of the two squares drawn on the legs of the original right triangle.

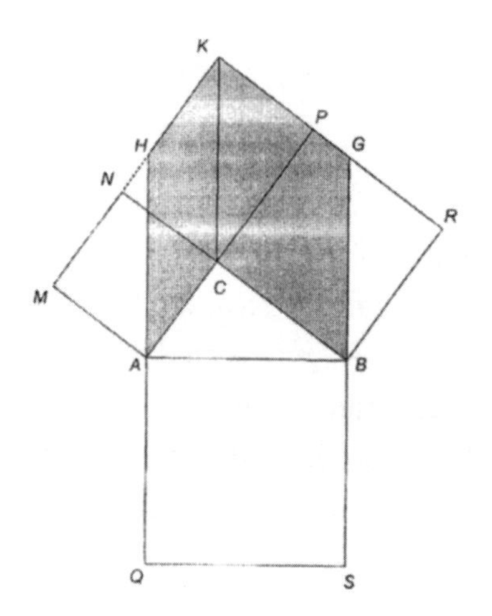


**Figure 2-7**

We can show that $\Delta GBR \cong \Delta ABC$, since $BC = BR$, $\angle GBR = \angle ABC$ (since each is composed of a right angle minus $\angle GBC$), and both triangles have a right angle. Therefore, $GB = AB$, and it follows that $GB = BS$.

Sliding this shaded figure down the length of $BS$ places it in the position shown in figure 2-8.

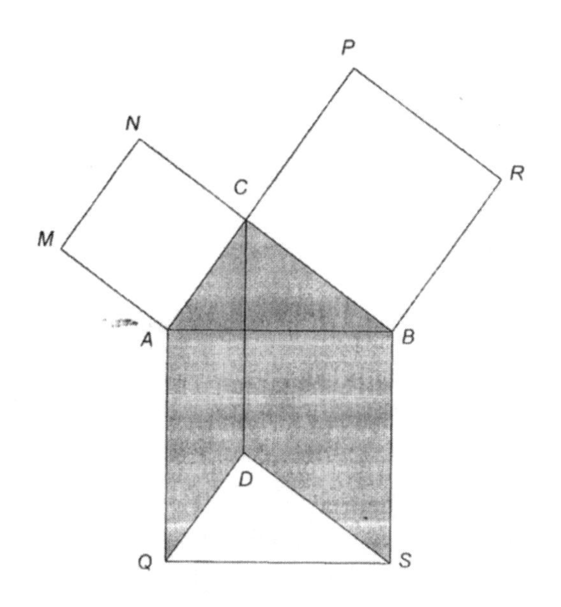


**Figure 2-8**

Since $\Delta ABC \cong \Delta QSD$, we can place the $\Delta ABC$ portion of the shaded region over $\Delta QSD$. This then gives us the area of square $ABSQ$. (See figure 2-9.) Thus, just by sliding some areas along established paths, we were able to prove that the area of square $ACNM$ + the area of square $BCPR$ = the area of square $ABSQ$, which is the Pythagorean Theorem.

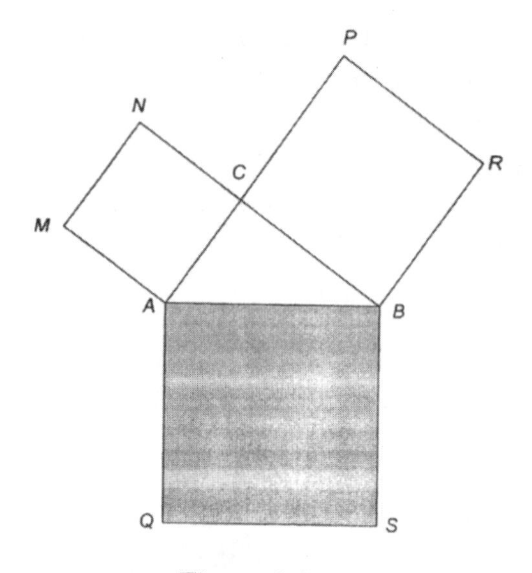


**Figure 2-9**

We can generalize the Pythagorean Theorem in two ways: to any nonright triangle (instead of a right triangle) and by using parallelograms (instead of squares on the sides). This was first discovered by Pappus of Alexandria in book IV of his *Mathematical Collection.* Consider the randomly drawn triangle $ABC$ in figure 2-10. On two of the sides ($AC$ and $BC$) parallelograms $ACNM$ and $BCPR$ are constructed. The sides $MN$ and $RP$ are extended to meet at point $K$. The segment $KC$ is extended so that $TY = KC$. Parallelogram $ABSQ$ is then drawn so that $AQ$ is equal and parallel to $TY$. Analogous to the proof of the Pythagorean Theorem, we can now demonstrate that the sum of the areas of parallelograms $ACNM$ and $BCPR$ equals the area of parallelogram $ABSQ$.

To do this, we use a similar argument as before. The equal areas of parallelograms:

$$ACNM = ACKX = ATYQ$$

Similarly, the following parallelograms also have equal areas:

$$BCPR = BCKZ = BTYS$$

The result then becomes obvious; namely,

> the area of parallelogram *ACNM* + the area of parallelogram *BCPR* = the area of parallelogram *BAQS*.

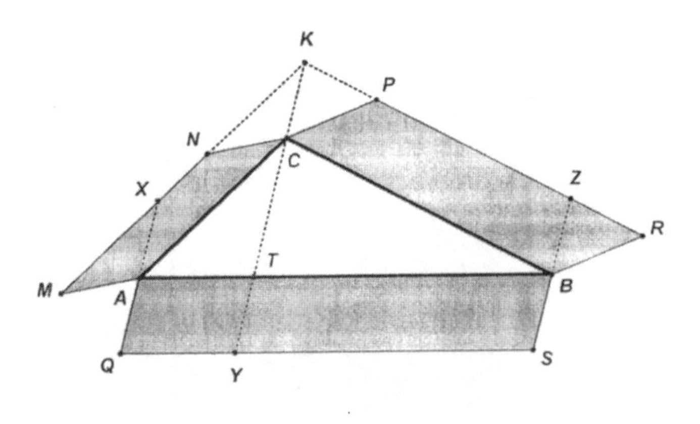


**Figure 2-10**

## Demonstration 3

Now that you might have developed some "dexterity" in comparing areas of triangles and quadrilaterals that share the same base and altitude (i.e., where the altitude's vertex lies on a line parallel to the base), we can embark on a nifty little excursion that will result in the Pythagorean Theorem. With our focus on right triangle *ABC*, we will begin by considering the two shaded triangles in figure 2-11.

We would like to show that the sum of the areas of the squares (*ACLK* and *BCED*) on the legs of right triangle *ABC* is equal to the area of the square (*ABFG*) on the hypotenuse. We begin by establishing that each of the two shaded triangles has half the area of the squares on the legs of right triangle *ABC*, respectively.[5] That is,

---

5. The justification of point *G*, lying on *KL* (figure 2-11), and point *F*, lying on the extension of *DE*, is embedded in the discussion of Demonstration 2 (pages 41–45).

$$\text{area } \Delta ACG = \frac{1}{2}\text{area } \square ACLK, \text{ and area } \Delta BCF = \frac{1}{2}\text{area } \square BCED$$

since in each case the triangle shares a base with the respective square and the altitude to that base is the same as that of the square.

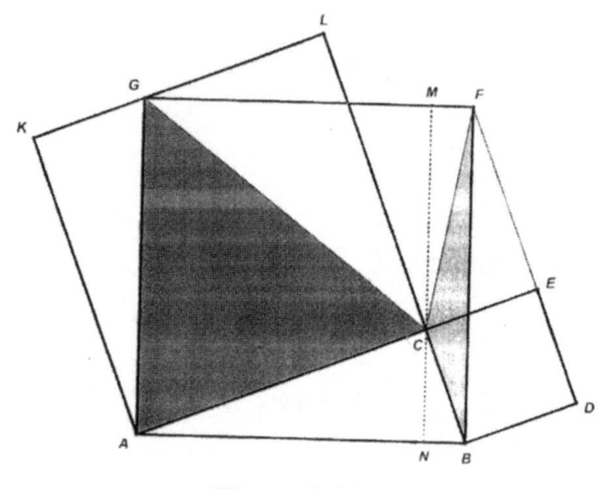


**Figure 2-11**

We must now show that the sum of the areas of these two shaded triangles is one-half the area of the square (*ABFG*) on the hypotenuse of right triangle *ABC*. We will consider the area of square *ABFG* in two parts: rectangles *ANMG* and *BNMF*. We can see that each of the two shaded triangles has half the area of one of these rectangles, that is,

$$\text{area } \Delta ACG = \frac{1}{2}\text{area } \square ANMG, \text{ and area } \Delta BCF = \frac{1}{2}\text{area } \square BNMF$$

Therefore, the sum of the areas of these two shaded triangles is equal to half the area of square *ABFG*. This implies that the sum of the areas of the squares (*ACLK* and *BCED*) on the legs of right triangle *ABC* is equal to the area of the square (*ABFG*) on the hypotenuse, since each is double the area of the same two shaded triangles. Thus, the Pythagorean Theorem has been proved!

## Demonstration 4

We can also prove the Pythagorean Theorem by cutting up parts to show the equal areas of the squares on the sides of the given right triangle.[6] Consider the partitioning of square *CBRP*, as shown in figure 2-12. By marking off equal segments[7] (*CE*, *PF*, *RG*, and *BH)* along the sides of square *CBRP*, we can create four shaded quadrilaterals that we can show to be congruent to each other and then can be placed into square *ABSQ* along with square *ACNM* to completely fill the square. You might want to try this demonstration with cardboard pieces. This again shows that the sum of the areas of the squares on the legs of a right triangle is equal to the area of the square on the hypotenuse. Again, the Pythagorean Theorem has been demonstrated.

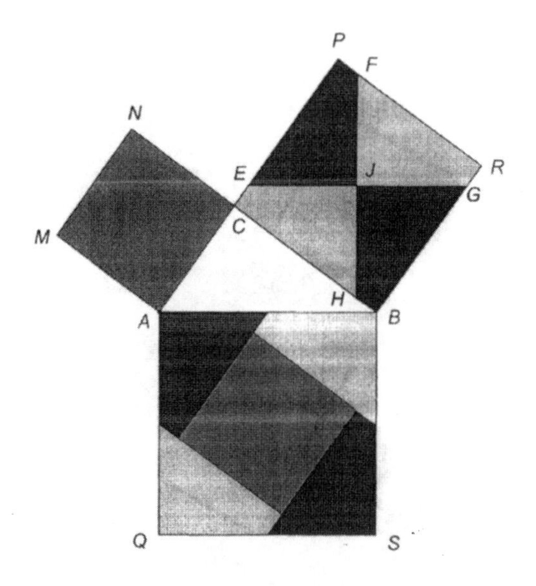


**Figure 2-12**

---

6. Roger B. Nelsen, *Proofs without Words* (Washington, DC: Mathematical Association of America, 1993), p. 6.

7. These equal segments are obtained by taking half the difference of the lengths of the legs of the right triangle.

## Demonstration 5

In figure 2-13 you will notice a black right triangle (which is our given right triangle *ABE*) embedded in the diagram. This demonstration[8] is attributed to Annairizi of Arabia (ca. 900 CE). The lighter-shaded and darker-shaded squares each have a respective leg-length of the given right triangle. Careful inspection will reveal that the dark-outlined square (*ABCD*) on the hypotenuse is composed of lighter-shaded and darker-shaded regions, which—if placed properly—would be exactly the area of the smaller and larger shaded squares. Therefore, the area of the square on the hypotenuse (*ABCD*) is equal to the sum of the areas of the squares on the legs of the given right triangle. Thus, once again we have demonstrated the truth of the Pythagorean Theorem, largely by inspection.

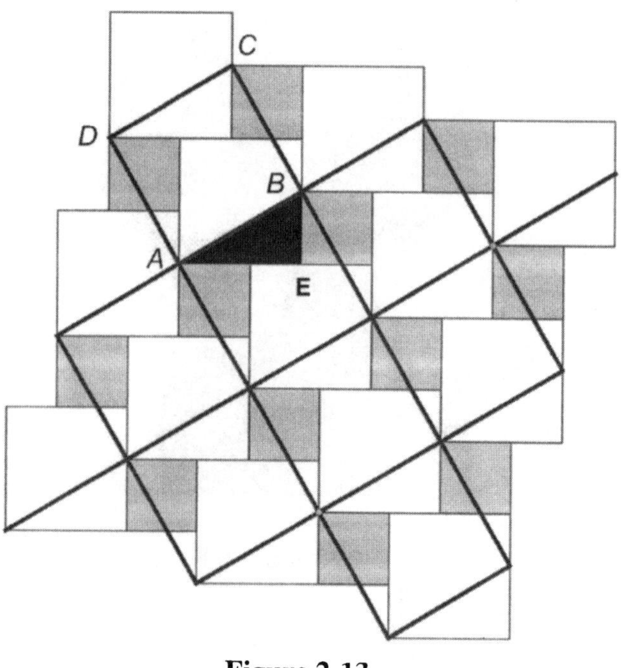


**Figure 2-13**

8. Roger B. Nelsen, *Proofs without Words II* (Washington, DC: Mathematical Association of America, 2000), p. 3.

## Demonstration 6

A similar technique was used by Liu Hui, a Chinese mathematician who lived in the third century. In the year 263 he published a book that provided solutions to the problems posed in *Nine Chapters on the Mathematical Arts*. His demonstration of the Pythagorean Theorem is shown in figure 2-14. You should be able to see how the component parts of squares *ACNM* and *ABSQ* can be made to perfectly fit into the square *BCPR*.[9] The symmetry of the figure and some clever location of complementary angles[10] will provide the justification of this construction. Thus, we can show that the area of square *ACNM* + the area of square *ABSQ* = the area of square *BCPR*, which is the Pythagorean Theorem.

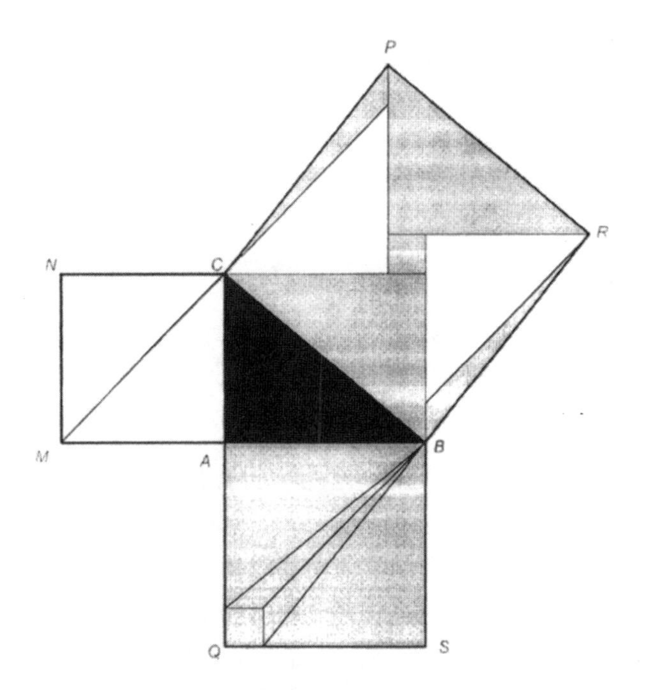


**Figure 2-14**

9. Philip D. Straffin Jr., "Liu Hui and the First Golden Age of Chinese Mathematics," *Mathematics Magazine* 71, no. 3 (June 1998): 170.

10. Complementary angles are two angles whose sum is 90°.

## Demonstration 7

In figure 2-15, we must focus on two pairs of unusual-looking quadrilaterals: one pair is *MNPR* and *TSBC* and the other pair is *MABR* and *CAQT*. Each of these pairs of quadrilaterals can be shown to be congruent. To do this you should recognize that the triangles *PCN*, *BCA*, and *QTS* are congruent.

Having accepted that, we can add the pairs of quadrilaterals to get equal areas (not congruence), for area *MNPR* + area *MABR* = area *TSBC* + area *CAQT*, which gives us area *MNPRBA* = area *TSBCAQ*.

By removing the overlapping triangle (actually our original right triangle *ABC*) from the two polygons *MNPRBA* and *TSBCAQ*, and then removing from each of the polygons the congruent triangles *PCN* and *QTS*, we get the area of square *ACNM* + the area of square *BCPR* = the area of square *ABSQ*, which is the Pythagorean Theorem.

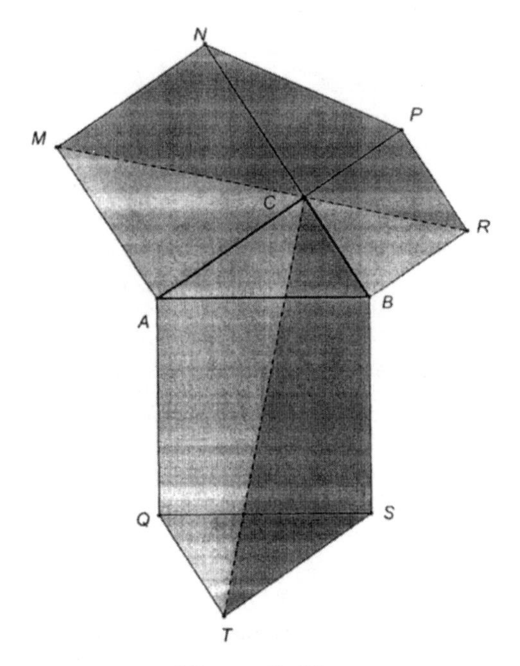


**Figure 2-15**

## Demonstration 8

Once again we have a visual demonstration of the Pythagorean Theorem. This time, the concern is how to construct the diagram[11] that you see in figure 2-16. We begin as usual with the right triangle *ABC* and the squares drawn on its three sides. Then we draw the common diagonal *MCR* of the two squares on the legs of the right triangle. The extensions of *QA* and *SB* intersecting with this diagonal give us points *U* and *V*, respectively. We then draw *NW* parallel to *AU*, and *PX* parallel to *BV*, where *W*, *U*, *X*, and *V* are on line *MR*.

The eight triangles into which the two smaller squares (*ACNM* and *BCPR*) are partitioned fit nicely and completely into the larger square so that we can conclude that the area of square *ACNM* + the area of square *BCPR* = the area of square *ABSQ*, which is the Pythagorean Theorem.

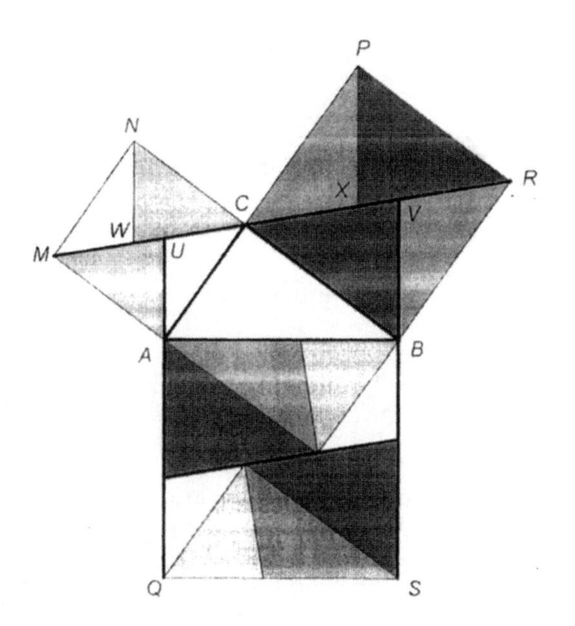


**Figure 2-16**

11. Elisha S. Loomis, *The Pythagorean Proposition* (Reston, VA: National Council of Teachers of Mathematics, 1968), p. 112.

## Demonstration 9

Rather than hint at the various steps that would constitute a proper formal deductive proof of the Pythagorean Theorem, we will do this one somewhat intuitively—which is easily supported by a deductive proof, if one so desires. Consider the diagram in figure 2-17.[12] We will make some sweeping generalizations, each of which can be proved, but for the sake of expediency we shall accept them directly.

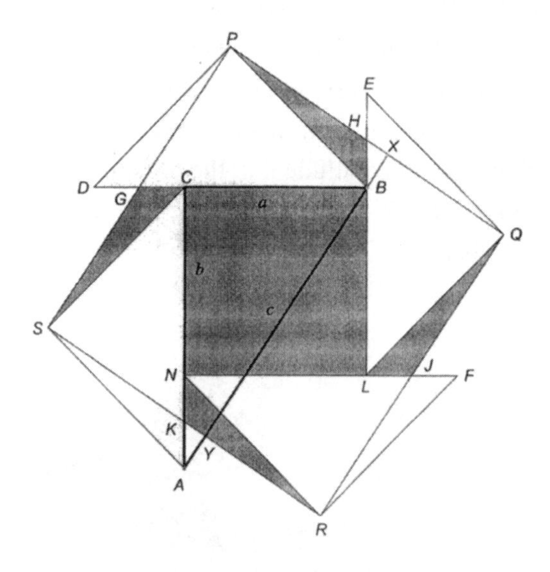


**Figure 2-17**

First, let's accept that the following triangles are all congruent to one another: $\Delta DGP$, $\Delta EHQ$, $\Delta FJR$, $\Delta AKS$, $\Delta CGS$, $\Delta HBP$, $\Delta LJQ$, and $\Delta NKR$.

Also the following larger triangles are isosceles right triangles and congruent to each other: $\Delta BPD$, $\Delta CSA$, $\Delta NRF$, and $\Delta LQE$.

By replacements of the smaller triangles, we can show that the sum of the areas of the four larger triangles $\Delta BPD$, $\Delta CSA$, $\Delta NRF$, and $\Delta LQE$, plus the area of the square *NLBC* is equal to the area of the square *PQRS*.

---

12. Poo-sung Park, *Mathematics Magazine* 72, no. 5 (December 1999): 407.

The four larger triangles, $\Delta BPD$, $\Delta CSA$, $\Delta NRF$, and $\Delta LQE$, when placed together, form a square with side length $AC$ (or $b$), which is one of the legs of the right triangle $ABC$.

We can show that, since $AY = BX$, the segment $XY$ (which is equal to the length of the side of square $PQRS$), is equal to $AB$, the hypotenuse of our focused-upon right triangle $ABC$.

We thus have essentially proved the Pythagorean Theorem by recalculating the sum of the areas of the squares on the legs of the right triangle $ABC$. On leg $AC$, the area of the square equals the sum of the areas of $\Delta BPD + \Delta CSA + \Delta NRF + \Delta LQE$, which is simply the area of a square with a side length $b$. That is, the area is $b^2$.

The area of the square $NLBC$ on leg $BC$ of the right triangle is $a^2$.

The sum of the areas of the squares on the legs of $AC$ and $BC$ is equal to the area of the largest square $PQRS$, which is $c^2$. Therefore, we have demonstrated the Pythagorean Theorem's truth, since we have shown that for right triangle $ABC$, $a^2 + b^2 = c^2$.

## Demonstration 10

The next visual proof of the Pythagorean Theorem is attributed to Leonardo da Vinci (1452–1519).[13] (Figure 2-18 shows Leonardo da Vinci's actual diagram.)

In book I (proposition 47) of his book *The Elements*, Euclid (ca. 300 BCE) popularized this proof of the Pythagorean Theorem. Although it may appear daunting, it is rather simple. Recall the relationship we described in figure 2-4b. Applying that result to figure 2-19, we find that the area of triangle $MAB$ is half the area of square $MACN$. Similarly, the area of triangle $CAQ$ is half the area of rectangle $AQHJ$. However, since $\Delta MAB \cong \Delta CAQ$,[14] we can

13. Eves, *Great Moments in Mathematics (Before 1650)*, pp. 29–31.

14. $MA = CA$, $QA = BA$, and $\angle MAB = \angle CAQ$. Then apply the SAS congruence postulate.

conclude that the area of square *MACN* is equal to the area of rectangle *AQHJ.*

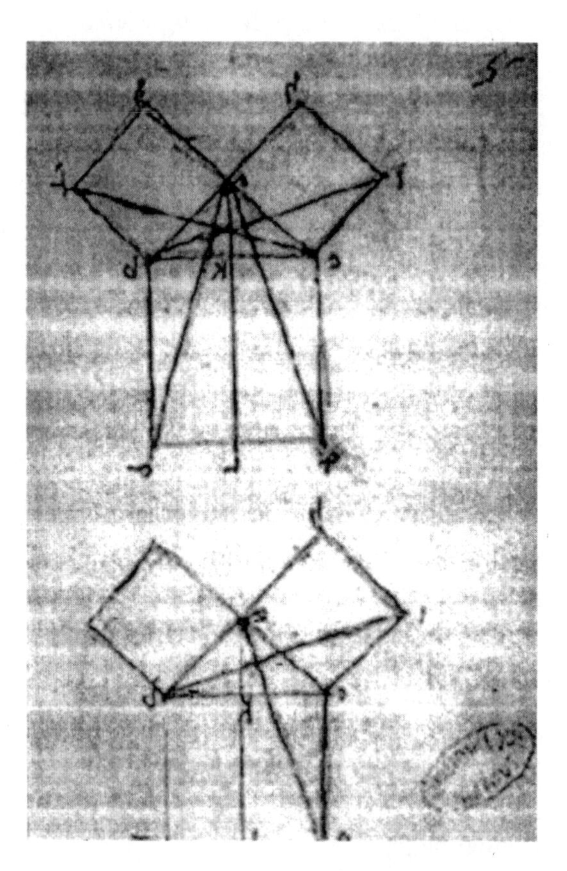

**Figure 2-18**

Using a similar strategy on the other side of this figure, we can show that square *PRBC* is equal in area to rectangle *BSHJ.* Therefore, by addition we get

The area of square *MACN* + the area of square *PRBC*
= the area of rectangle *AQHJ* + the area of rectangle *BSHJ*
= the area of square *ABSQ*

which again justifies the Pythagorean Theorem.

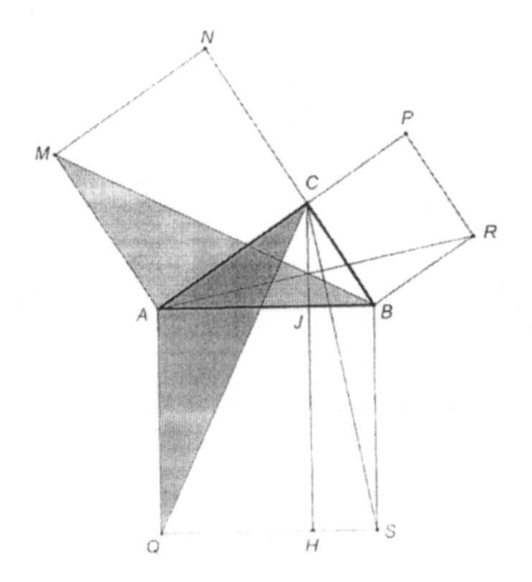


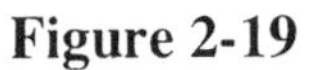
**Figure 2-19**

You can observe this in stages with the simulated motions shown in figure 2-20.

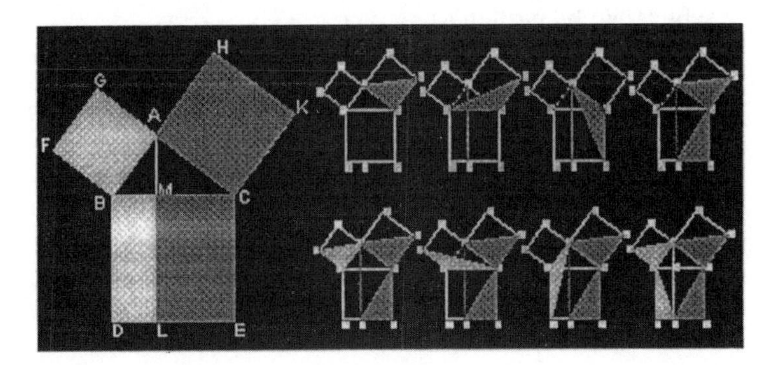


**Figure 2-20**

## Demonstration 11

The traditional proof of the Pythagorean Theorem, one that is found in most geometry textbooks, builds on the relationships of three similar triangles as embedded in figure 2-21.

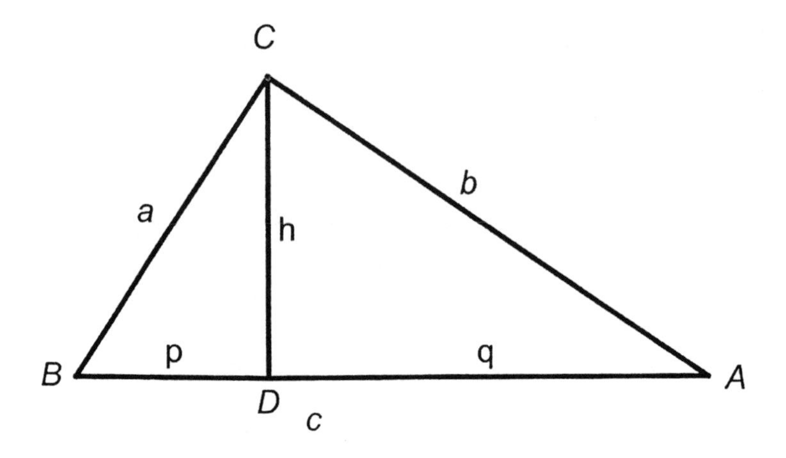


**Figure 2-21**

In figure 2-21 we have $\Delta ABC \sim \Delta BCD \sim \Delta CAD$.[15] Consequently, we get the following proportions:

From $\Delta ABC \sim \Delta BCD$, we get $\frac{c}{a} = \frac{a}{p}$,

which can be written as $a^2 = cp$

From $\Delta ABC \sim \Delta CAD$, we get $\frac{c}{b} = \frac{b}{q}$,

which can be written as $b^2 = cq$

By adding these two equations, we get

$$a^2 + b^2 = cp + cq = c(p+q)$$

However, $p+q=c$

Therefore, $a^2 + b^2 = c^2$, and the Pythagorean Theorem is once again established.

15. $\angle BCD$ is complementary to $\angle ACD$, as is $\angle A$. Therefore, $\angle BCD = \angle A$, so we have, along with the right angle of each of the three triangles, similarity established.

## Demonstration 12

Some demonstrations of the Pythagorean Theorem have been done by people not known for their mathematical prowess. In 1876, while still a member of the House of Representatives, the soon-to-be twentieth president of the United States, James A. Garfield, produced the following proof.[16] Garfield was previously a professor of classics and has the distinction today of being the only sitting member of the House of Representatives to have been elected president of the United States. Let's take a look at the proof he discovered.

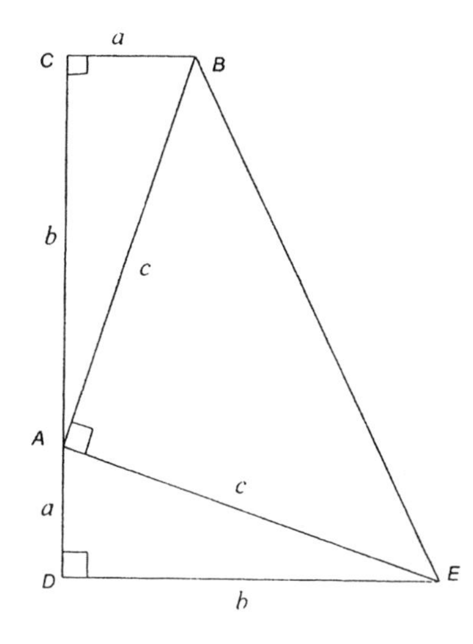


**Figure 2-22**

In figure 2-22, $\Delta ABC \cong \Delta EAD$, and all three triangles in the diagram are right triangles.[17] Recall that the area of the trapezoid[18]

16. James A. Garfield, "Pons Asinorum," *New England Journal of Education* 3, no. 161 (1876).

17. Triangle $ABE$ is a right triangle because $\angle BCA$ and $\angle EAD = 90°$. Therefore, $\angle BAE$ is a right angle.

18. A *trapezoid* is a quadrilateral with exactly one pair of opposite sides parallel.

*DCBE* is half the product of the altitude $(a + b)$ and the sum of the bases $(a + b)$, which we can write as $\frac{1}{2}(a+b)^2$.

We can also get the area of the trapezoid *DCBE* by finding the sum of the areas of each of the three right triangles:

$$\frac{1}{2}ab+\frac{1}{2}ab+\frac{1}{2}c^2$$
$$=2\left(\frac{1}{2}ab\right)+\frac{1}{2}c^2$$

We can then equate the two expressions that represent the area of the entire trapezoid:

$$2\left(\frac{1}{2}ab\right)+\frac{1}{2}c^2=\frac{1}{2}(a+b)^2$$

This can be simplified to

$$2ab+c^2=(a+b)^2$$
$$2ab+c^2=a^2+2ab+b^2$$
$$c^2=a^2+b^2$$

This is the Pythagorean Theorem as applied to right triangle *ABC*.

An astute reader may notice that Garfield's proof is somewhat similar to the one believed to have been used by Pythagoras. If we "complete" a square from the given trapezoid in figure 2-22, we get a configuration similar to that shown in figure 2-23.

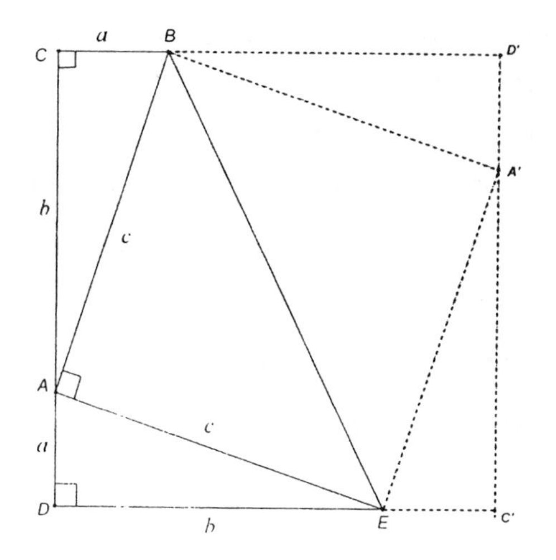


**Figure 2-23**

A high school student[19] manipulated Garfield's configuration and came up with yet another proof of the Pythagorean Theorem. She flipped[20] figure 2-22 to get the figure shown in figure 2-24.

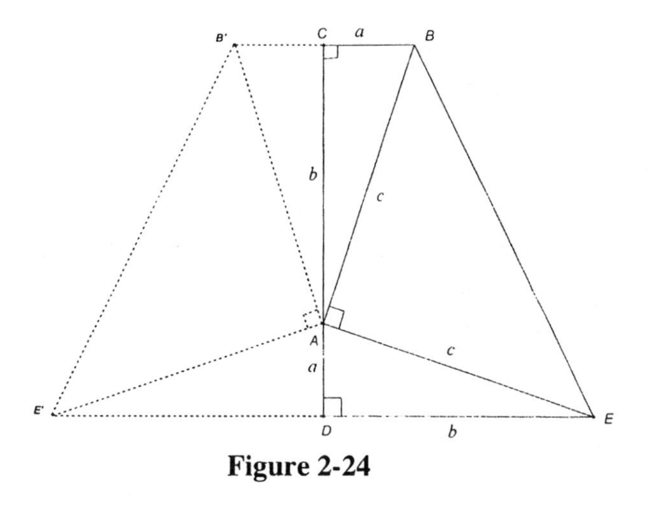


**Figure 2-24**

19. Jamie deLemos, a student at Framingham High School in Framingham, Massachusetts, presented her proof in *Mathematics Teacher* 88, no. 1 (January 1995): 79.

20. This is also called a reflection in line *CAD*.

She simply found the area of the trapezoid as the sum of the areas of the various right triangles. First, the area of the trapezoid $BB'E'E$ is equal to one-half the product of its altitude and the sum of its bases, namely, $\frac{1}{2}(a+b)(2a+2b)$. Now the area of this trapezoid is also the sum of the six right triangles that comprise it. That is, four times the area of triangle $ABC$ and twice the area of triangle $ABE$, that is, $4\left(\frac{1}{2}ab\right)+2\left(\frac{1}{2}c^2\right)$.

Equating these two expressions for the same area (trapezoid $BB'E'E$), we get

$$\frac{1}{2}(a+b)(2a+2b)=4\left(\frac{1}{2}ab\right)+2\left(\frac{1}{2}c^2\right)$$

Then simplifying, we get

$$(a+b)^2=2ab+c^2$$
$$a^2+2ab+b^2=2ab+c^2$$
$$a^2+b^2=c^2$$

As the student herself says: "Who would have guessed that an average geometry student could come up with an original derivation[21] of this famous theorem? Logic and creativity are the keys, plus a little push and some thinking." This might serve to motivate the reader to seek other proofs of the Pythagorean Theorem.

## Demonstration 13

From point $P$ on the circle with center $O$, and radius length $c$, we draw a perpendicular to the diameter $AB$. (See figure 2-25.)

---

21. Although this appears to be another method of proving the Pythagorean Theorem, it is actually another version of Garfield's proof.

Since we can show that $\Delta PCB \sim \Delta ACP$, we can then set up the following proportion: $\frac{b}{c-a}=\frac{c+a}{b}$.

This can then be transformed to get

$$(c-a)(c+a)=b^2$$
$$c^2-a^2=b^2$$
$$c^2=a^2+b^2$$

By now you should recognize the Pythagorean Theorem!

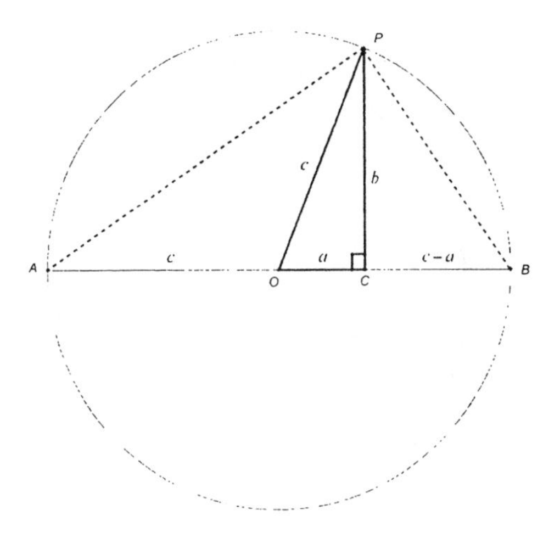


**Figure 2-25**

## Demonstration 14

Perhaps the most famous theorem involving cyclic quadrilaterals (i.e., quadrilaterals that can be inscribed in a circle)[22] is that attributed to Claudius Ptolemaeus of Alexandria (popularly known as

22. Remember, not all quadrilaterals can be inscribed in a circle. For example, the only parallelogram that can be inscribed in a circle is either a square or a rectangle.

Ptolemy). In his major astronomical work, the *Almagest* (ca. 150 CE), he states this theorem on cyclic quadrilaterals as follows:

> The product of the lengths of the diagonals of a cyclic quadrilateral equals the sum of the products of the lengths of the pairs of opposite sides. (Ptolemy's Theorem)[23]

In figure 2-26, Ptolemy's Theorem gives us the following relationship: $AC \cdot BD = AB \cdot CD + AD \cdot BC$ (proof is provided in appendix A).

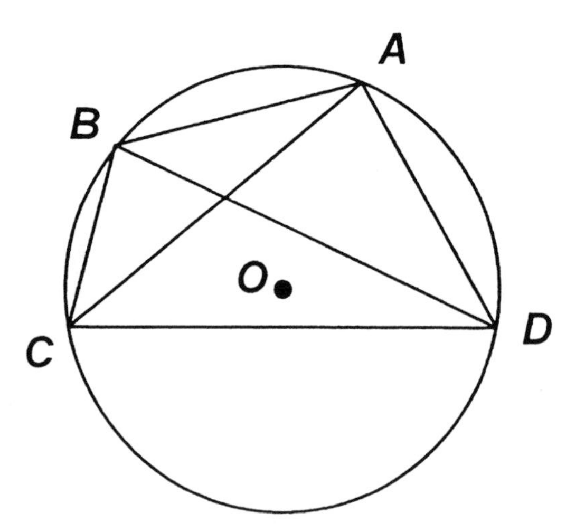


**Figure 2-26**

If we apply Ptolemy's Theorem to the rectangle *ABCD*—which is always inscribable in a circle—we get (for figure 2-27) the following: $AC \cdot BD = AB \cdot CD + AD \cdot BC$.

23. The Greek title *Syntaxis Mathematica* means mathematical (or astronomical) compilation. The Arabic title *Almagest* is a renaming meaning "great collection (or compilation)." The book is a manual of all the mathematical astronomy that the ancients knew at that time. Book I of the thirteen books that are in this monumental work contains the theorem (6.11) that now bears Ptolemy's name.

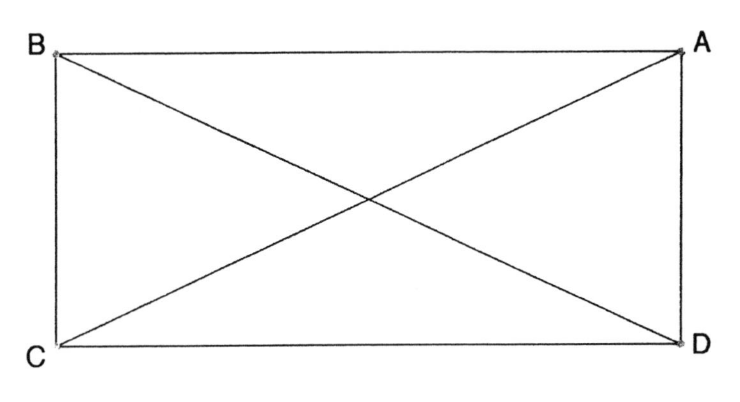


**Figure 2-27**

However, for rectangle $ABCD$,

$$AB = CD,\ AD = BC, \text{ and } AC = BD$$

Making the proper substitutions we get $AB^2 + BC^2 = AC^2$, which is the Pythagorean Theorem as applied to triangle $ABC$. Thus, we have used Ptolemy's Theorem to prove the Pythagorean Theorem.

## Demonstration 15

We can reach back to our study of high school geometry and recall a very simple theorem, namely, that when two chords[24] intersect in the circle, the product of the segments of one chord is equal to the product of the segments of the other chord.

In figure 2-28 this would mean that $pq = rs$.

24. A *chord* is a line segment joining two points on a circle.

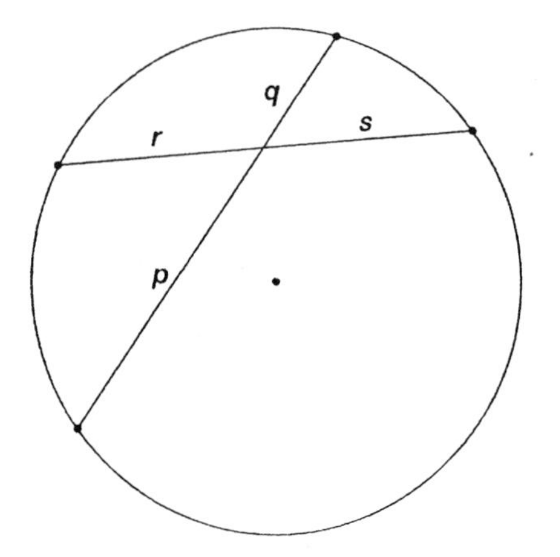


**Figure 2-28**

Let us now consider the circle with diameter *AB* perpendicular to chord *CD* (see figure 2-29).

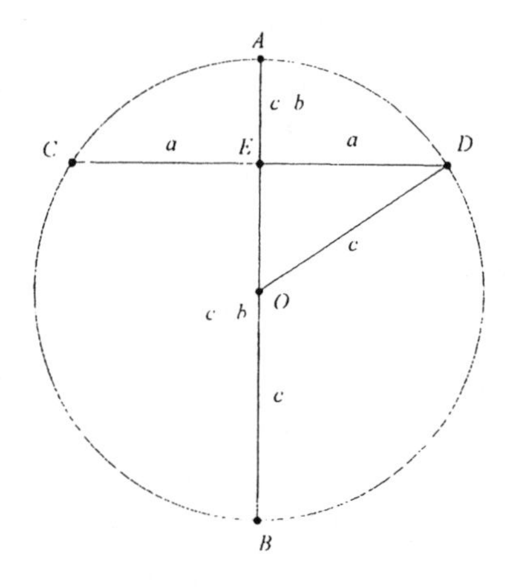


**Figure 2-29**

From the theorem stated above and where $AE = c - b$ and $BE = c + b$, we get $(c-b)(c+b) = a^2$.

Then $c^2 - b^2 = a^2$, and therefore, $a^2 + b^2 = c^2$. The Pythagorean Theorem is proved again.

## Demonstration 16

Analogous to demonstration 15 we shall again go back to high school geometry for a theorem. It states that when a tangent[25] and a secant[26] are drawn to a circle from the same external point, the tangent is the mean proportional[27] between the secant and its external segment.

In figure 2-30, this tells us that $\frac{BD}{BC} = \frac{BC}{BE}$.

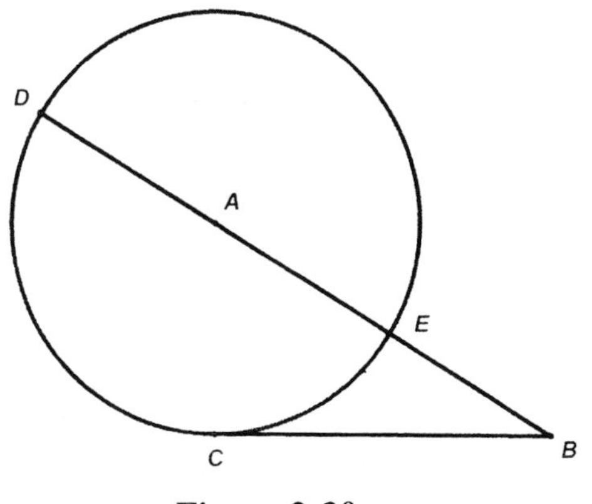


**Figure 2-30**

Let's apply this to figure 2-31, where we have right triangle $ABC$, where $BC = a$, $AC = b$, and $AB = c$. We get $\frac{c+b}{a} = \frac{a}{c-b}$.

Then cross-multiplying:

$$a^2 = (c+b)(c-b) = c^2 - b^2$$

25. A line from a point outside the circle that touches the circle in exactly one point is called a *tangent* to the circle.

26. A line from a point outside the circle that touches the circle in exactly two points is called a *secant* to the circle.

27. The mean proportional is the quantity in the means position of a proportion. In the proportion $\frac{a}{x} = \frac{x}{b}$, $x$ is called the mean proportional.

Then $a^2 + b^2 = c^2$, which is the Pythagorean Theorem applied to triangle *ABC*.

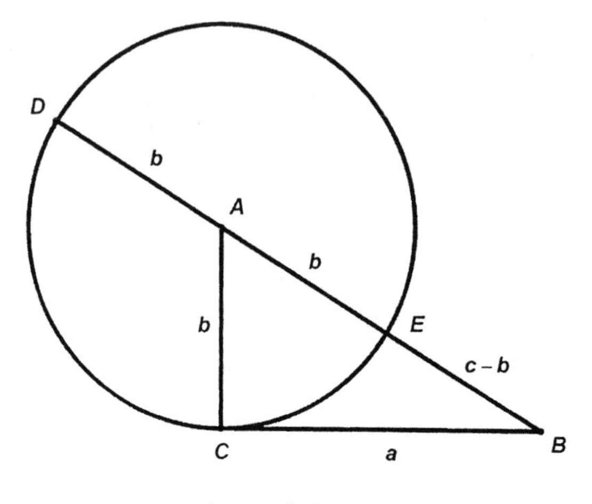


**Figure 2-31**

## Demonstration 17

A well-known formula for finding the area of a triangle is Heron's formula, which is named after the Greek mathematician Heron of Alexandria (20–62 CE). Typically, to find the area of a triangle we need to know the lengths of the triangle's altitude and its base. Yet Heron's formula allows us to find the area of a triangle when all we know are the lengths of the three sides of the triangle, and not its altitude. If we know the lengths of the three sides, then we also know the perimeter of the triangle. Hence we can find the half of the perimeter (figure 2-32), which we call the semiperimeter, $s = \frac{a+b+c}{2}$.

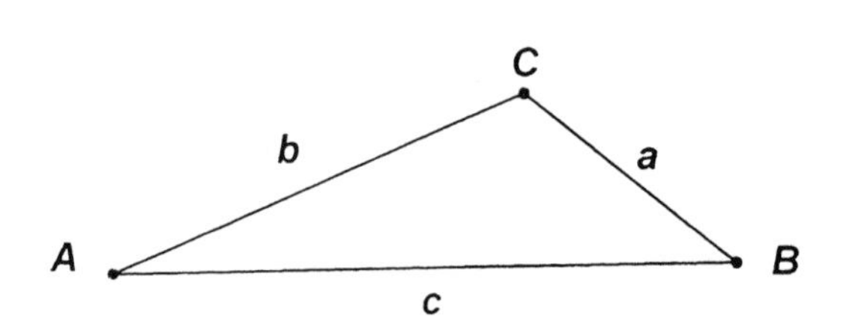


**Figure 2-32**

Heron's formula for finding the area of a triangle of side lengths $a$, $b$, and $c$ is: $A = \sqrt{s(s-a)(s-b)(s-c)}$.

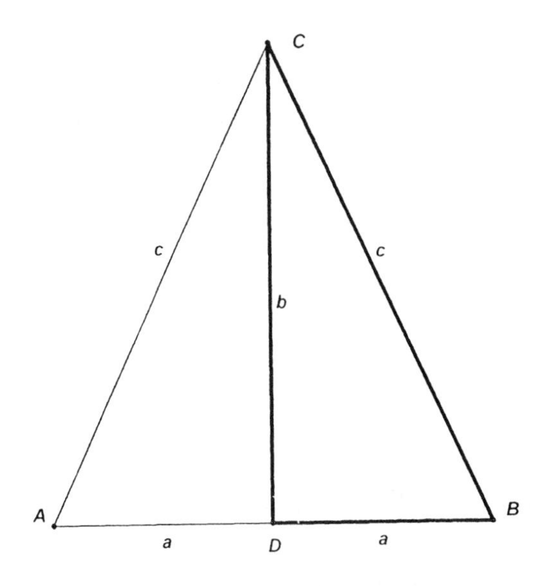


**Figure 2-33**

In figure 2-33 we would like to show that for right triangle $DBC$, the Pythagorean Theorem holds true. To do this we will apply Heron's formula to the isosceles triangle $ABC$ as follows:

$$\text{area of } \Delta ABC = \sqrt{(c+a)(c+a-c)(c+a-c)(c+a-2a)},$$

$$\text{where semiperimeter } s = \frac{2c+2a}{2} = c+a$$

However, the traditional formula for finding the area of triangle $ABC$ (using the triangle's base and altitude) is $\frac{1}{2}(2ab) = ab$.

Equating these two area formulas for triangle $ABC$—and using just a little elementary algebra—will give us the Pythagorean Theorem as applied to triangle $DBC$.

$$\text{area of } \Delta ABC = \sqrt{(c+a)(c+a-c)(c+a-c)(c+a-2a)} = ab$$
$$= \sqrt{(c+a)(a)(a)(c-a)} = ab$$
$$= \sqrt{(c^2-a^2)(a^2)} = ab$$

By squaring both sides of this equation we get

$$(c^2-a^2)(a^2) = a^2b^2$$
$$c^2 - a^2 = b^2$$
$$a^2 + b^2 = c^2$$

which establishes the Pythagorean Theorem!

## Demonstration 18

The only limit to discovering a new proof of the Pythagorean Theorem is one's imagination. By simply placing two congruent right triangles in a particular configuration, we can come up with a variety of proofs, or demonstrations, of the Pythagorean Theorem. We will show three different such demonstrations and leave the reader to find others from this placement of triangles.

Consider the two congruent right triangles *ABC* and *CDG* as shown in figure 2-34. They share point *C* for a vertex and point *G* lies on side *BC*.

This time our discussion of this unusual configuration of two congruent right triangles will require a few more words than we have bargained for in the spirit of this chapter. However, we will be using nothing more than some of the very basics of high school geometry and we will cover it as thoroughly as possible. As we begin, recall that $\Delta ABC \cong \Delta CDG$. Since $\angle B = \angle D$ and the vertical angles $\angle BEG = \angle DEF$, it follows that $\Delta BEG \sim \Delta DEF$. And thus, $\angle DFE = 90°$.

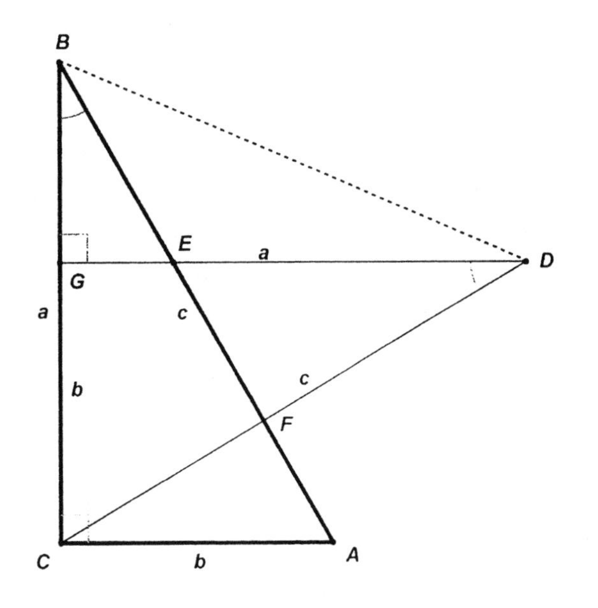


**Figure 2-34**

We will now make a short digression to find the length of $BF$ in terms of $b$ and $c$. Notice that $\Delta AFC \sim \Delta ACB$. Whereupon, it follows that $\frac{AB}{AC} = \frac{AC}{AF}$. By substituting the noted values[28] shown in figure 2-34, we get $\frac{c}{b} = \frac{b}{AF}$, which then allows us to have $AF = \frac{b^2}{c}$.

We then have $BF = AB - AF = c - \frac{b^2}{c} = \frac{c^2 - b^2}{c}$.

We will now see the reason for setting up this unusual configuration shown in figure 2-34. We can get the area of triangle $DBC$ in two different ways:

Once with $CD$ as the base and $BF$ as the altitude:

$$\text{area } \Delta DBC = \frac{1}{2}(CD)(BF) = \frac{1}{2}(c)\left(\frac{c^2 - b^2}{c}\right) = \frac{1}{2}(c^2 - b^2)$$

28. In figure 2-34, $BC = GD = a$, $AC = GC = b$, and $AB = CD = c$.

And once with $BC$ as the base and $DG$ as the altitude:

$$\text{area } \Delta DBC = \frac{1}{2}(BC)(DG) = \frac{1}{2}(a)(a) = \frac{1}{2}a^2$$

When we equate these two expressions for the same area, we get $c^2 - b^2 = a^2$, or $a^2 + b^2 = c^2$, which, of course, is our familiar Pythagorean Theorem.

## Demonstration 19

Building from these relationships, we can also prove the Pythagorean Theorem by considering the quadrilateral *ACBD*, as shown in figure 2-35. Here we will establish the area of quadrilateral *ACBD* in two different ways, just as we established the area of triangle *DBC* in two different ways in demonstration 18.

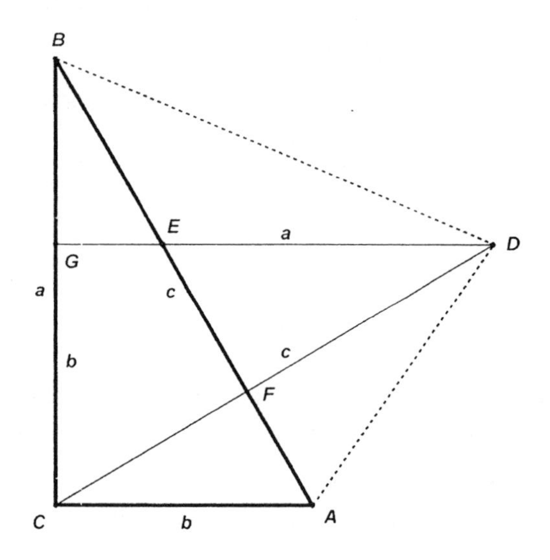


**Figure 2-35**

$$\begin{aligned} \text{area } ACBD &= \text{area } \Delta DBC + \text{area } \Delta CAD \\ &= \frac{1}{2}(CD)(BF) + \frac{1}{2}(CD)(AF) \\ &= \frac{1}{2}(CD)(BF + AF) \\ &= \frac{1}{2}(CD)(AB) \\ &= \frac{1}{2}c^2 \end{aligned}$$

Since $GC = b$ and $AC = b$, using $AC$ as the base and $GC$ as the height, we get the area of $\Delta CAD = \frac{1}{2} GC \cdot AC = \frac{1}{2}b^2$

Thus, the area of quadrilateral $ACBD$ is the sum of the areas of the two triangles, $DBC$ and $CAD$, namely, $\frac{1}{2}a^2 + \frac{1}{2}b^2$.

However, earlier we found the area of quadrilateral $ACBD$ to be $\frac{1}{2}c^2$. By equating these two expressions for the same area (that of quadrilateral $ACBD$) we get

$$\frac{1}{2}a^2 + \frac{1}{2}b^2 = \frac{1}{2}c^2, \text{ or } a^2 + b^2 = c^2$$

Again we proved the Pythagorean Theorem from this unusual configuration.

## Demonstration 20

Again we will use the same configuration of the two congruent right triangles that we used in the previous two demonstrations. However, this time we will add another line segment to the figure to form a square $ACGH$ (see figure 2-36).

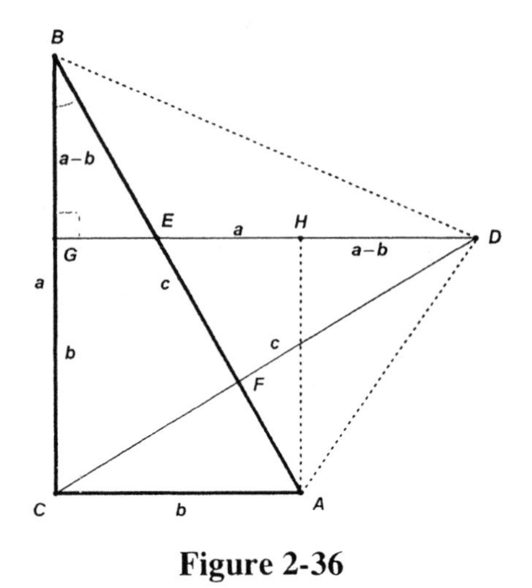


**Figure 2-36**

In the previous demonstration, we established that the area $ACBD = \frac{1}{2}c^2$. However, as before, we can also express the area of this quadrilateral as the sum of the areas of square $ACGH$, triangle $AHD$ and triangle $BGD$. That is,

$$\text{area } \square ACGH = b^2, \text{ area } \Delta AHD = \frac{1}{2}b(a-b) = \frac{ab-b^2}{2},$$

$$\text{and area } \Delta BGD = \frac{1}{2}a(a-b) = \frac{a^2-ab}{2}$$

Now summing these three areas we get

$$b^2 + \frac{ab-b^2}{2} + \frac{a^2-ab}{2} = \frac{2b^2}{2} + \frac{ab-b^2}{2} + \frac{a^2-ab}{2} = \frac{a^2+b^2}{2}$$

But remember that we also found that area $ACBD = \frac{1}{2}c^2$.

Therefore, $\frac{c^2}{2} = \frac{a^2+b^2}{2}$, or $a^2 + b^2 = c^2$.

And so we used essentially the same unusual configuration to prove the Pythagorean Theorem three different ways, yet we used the same idea of equating areas.

## Demonstration 21

There are proofs of the Pythagorean Theorem that show some real imagination or ingenuity. As our last proof, we provide just such a curious approach. It is actually quite simple, but uses a rather unexpected path. We begin with right triangle *ABC* and mark off point *H* on *BC*, so that $CH = AC = b$. We then extend *AC* to point *D*, so that $CD = CB = a$. When we draw *AH* and extend it to meet *BD* at *E*, we find that triangle *AED* is an isosceles right triangle, since triangle *ACH* is an isosceles right triangle and both triangles share the $\angle CAH = 45°$. Figure 2-37 shows this configuration with *DH* extended to meet *AB* at point *F*. We note that $\Delta ACB \cong \Delta HCD$.[29] Thus, $DH = c$, and for convenience we shall let $FH = x$.

We know that the altitudes of a triangle are concurrent. Therefore, since point *H* is the intersection of two of the altitudes (*BC* and AE) of triangle *ABD*, it follows that *DF* must also be an altitude of the triangle.

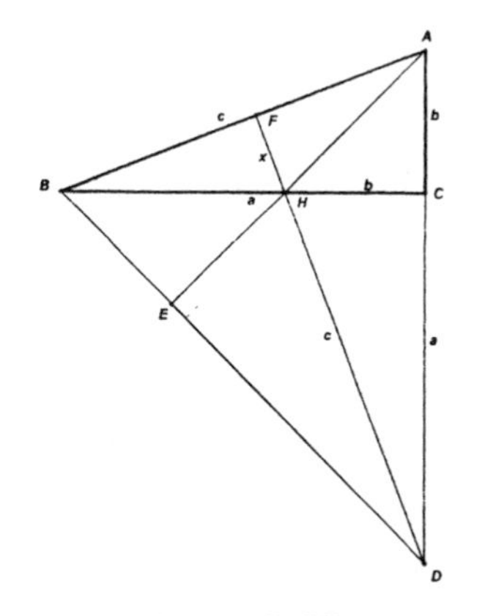


**Figure 2-37**

29. The congruence is established since the two legs of each of these right triangles are correspondingly congruent.

Now that we have set up this rather strange "arrangement," we are ready to establish the Pythagorean Theorem for right triangle *ABC*. To do this, we will express the area of triangle *ABD* in two different ways:

$$\text{area } \Delta ABD = \frac{1}{2}c\left(c + x\right)$$

$$\begin{aligned}\text{area } \Delta ABD &= \text{area } \Delta AHB + \text{area } \Delta AHC + \text{area } \Delta BCD \\ &= \frac{1}{2}cx + \frac{1}{2}b^2 + \frac{1}{2}a^2\end{aligned}$$

Now equating these two expressions of the area of triangle *ABD*:

$$\begin{aligned}\frac{1}{2}c\left(c+x\right) &= \frac{1}{2}cx + \frac{1}{2}b^2 + \frac{1}{2}a^2 \\ c^2 + cx &= cx + b^2 + a^2 \\ a^2 + b^2 &= c^2\end{aligned}$$

Thus, we established the Pythagorean Theorem from a rather unexpected and strange configuration.

※※※

After all the elegant and unusual demonstrations (or proofs) of the Pythagorean Theorem, we ought to be able to justify the converse of the Pythagorean Theorem. That is, that *if a triangle has side lengths* a, b, *and* c, *and the relationship that* $a^2 + b^2 = c^2$ *is true, then the angle between the shorter two sides,* a *and* b, *must be a right angle*. We will explore this shortly.

There are well over four hundred such demonstrations of the Pythagorean Theorem published to date and many more will likely be discovered in years to come. It is often said that the number of proofs—both geometric and algebraic—is boundless. Perhaps some of the proofs we have shown will stimulate the reader to find others—maybe even an original one! However, for our purposes we exhibited some of the more interesting proofs—both historically and esthetically—and if we have motivated some to search for other proofs, all the better.

# Chapter 3

# Applications of the Pythagorean Theorem

There are boundless applications of the Pythagorean Theorem. Often, we use it without a conscious attribution to it. A simple illustration of the use of the Pythagorean Theorem can be seen from the following example. Suppose you were about to buy a circular tabletop and wanted to know if it will fit through the doorway of your house, which measures 36 inches wide and 80 inches high. The tabletop that you are contemplating buying has a diameter of 88 inches. Will it fit through the doorway? Well, its diameter is clearly longer than the door's height.

With the help of the Pythagorean Theorem we can determine what the diagonal length of the doorway is and then see how that compares with the diameter of the tabletop.[1]

The diagonal of the doorway =

$$\sqrt{36^2 + 80^2} = \sqrt{1{,}296 + 6{,}400} = \sqrt{7{,}696} \approx 87.727$$

1. When we use the Pythagorean Theorem ($a^2 + b^2 = c^2$) to find the hypotenuse length, we convert it to the form: $c = \sqrt{a^2 + b^2}$.

**Figure 3-1**

Because this diagonal length is less than the diameter of the tabletop, we can conclude that the tabletop will not fit through the door. Our knowledge of the Pythagorean Theorem can help us avoid a frustrating situation: buying a tabletop that you cannot bring into your house.

## Constructing Irrational-Number Lengths

The Pythagorean Theorem led ancient mathematicians to a dilemma, namely, *irrational* numbers—that is, numbers that are not rational, which means they cannot be expressed as the *ratio* of two integers (i.e., as a common fraction). Examples of numbers that are irrational are $\sqrt{2}$, $\sqrt{5}$, $\sqrt{41}$, $e$, and $\pi$.[2] The dilemma then arises: how can you draw a line of length $\sqrt{7}$ inches long, when the rulers we use do not indicate such inch-subdivisions?[3] Once again the Pythagorean Theorem comes in handy. There are many ways of

---

2. For more about π, see Alfred S. Posamentier and Ingmar Lehmann, *π: A Biography of the World's Most Mysterious Number* (Amherst, NY: Prometheus Books, 2004).

3. We say that the irrational lengths are *incommensurable* with the rational number lengths.

constructing a length $\sqrt{7}$ inches long. To do this, we will employ the Pythagorean Theorem.

We begin with isosceles right triangle $BAO$, where $AB$ and $AO$ are each of length 1. (See figure 3-2.) By applying the Pythagorean Theorem: $1^2 + 1^2 = BO^2$, we then get $BO = \sqrt{2}$. We then proceed to construct $BC$ of length 1 to be perpendicular to $BO$. Once again applying the Pythagorean Theorem, we find $CO = \sqrt{3}$ (since $CO = \sqrt{1^2 + (\sqrt{2})^2} = \sqrt{1+2} = \sqrt{3}$). Continuing this process we get $DO = \sqrt{4}$, $EO = \sqrt{5}$, $FO = \sqrt{6}$, and $GO = \sqrt{7}$, which is what we originally sought.[4] We can continue this process and get further irrational lengths as shown in figure 3-2.

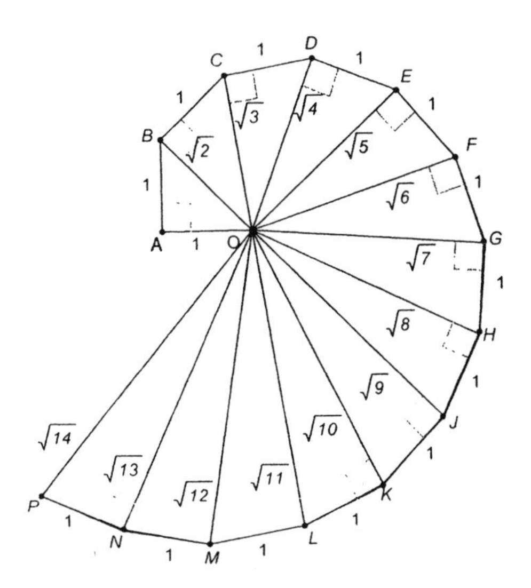


**Figure 3-2**

4. It is also possible to construct a length $\sqrt{x}$, by using the following construction:

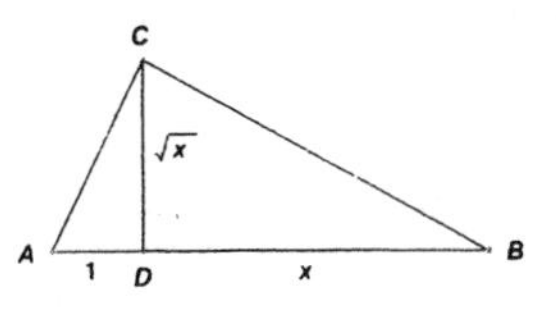

## The Converse of the Pythagorean Theorem—Establishing a Right Angle

The obvious may have passed us by. We can use the Pythagorean Theorem to determine if an angle of a triangle is a right angle. In order for us to use the converse of the Pythagorean Theorem, we must first establish its truth—that is, prove it is true. This can be done in a rather simple—or subtle—manner.

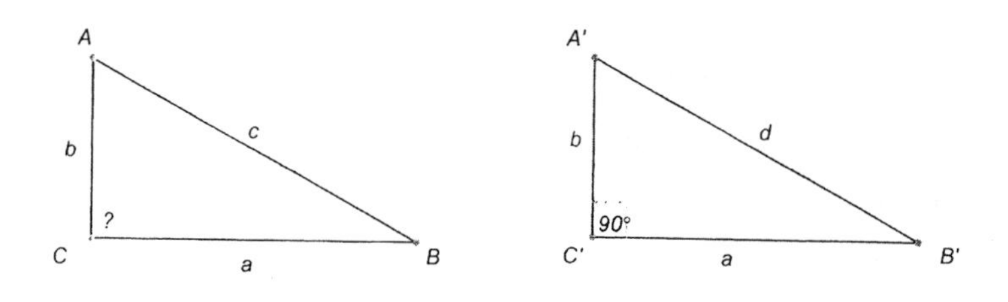


**Figure 3-3**

In figure 3-3, we have triangle *ABC* with side lengths *a*, *b*, and *c*, and triangle *A'B'C'* with side lengths *a*, *b*, and *d*. We are told that for triangle *ABC*, $a^2 + b^2 = c^2$. Our task is to show that for triangle *ABC*, angle *C* is a right angle.

To do this we will use right triangle *A'B'C'*, for which we know, by the Pythagorean Theorem, that $a^2 + b^2 = d^2$. Therefore, $d^2 = c^2$, or $d = c$. This makes $\Delta ABC \cong \Delta A'B'C'$, since the corresponding sides are equal in length. Therefore, $\angle C = \angle C' = 90°$. This, then, proves that if for a triangle, $a^2 + b^2 = c^2$, then the triangle must be a right triangle. This is one of the most important applications of the Pythagorean Theorem. It was employed as far back as the ancient Egyptians, who, as we mentioned in chapter 1, used rope stretchers to construct a right angle. Recall that they took a rope and placed 12 knots equally spaced along it. By stretching the rope to form a triangle of side lengths 3, 4, and 5, they obtained a right angle between the two shorter sides. (See figure 3-4.)

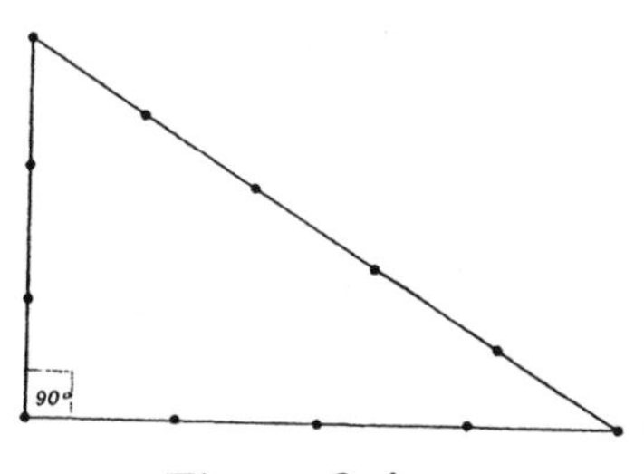


**Figure 3-4**

## Determining If an Angle Is Obtuse or Acute

Just as we can use the Pythagorean Theorem to determine if an angle is a right angle, it is also useful in determining if an angle of a triangle is acute (less than 90°) or obtuse (greater than 90°). In short, if $a^2 + b^2 < c^2$, then the angle between the sides of length $a$ and $b$ is obtuse. And if $a^2 + b^2 > c^2$, then the angle between the sides of length $a$ and $b$ is acute. The proof of this may be found in appendix A.

## The Lunes and the Triangle

The area of a circle is not typically commensurate with the areas of rectilinear figures. That is, it is quite unusual to be able to construct a circle equal in area to a rectangle, or a parallelogram, or for that matter any other figure composed of straight lines, which we refer to as "rectilinear" figures. However, with the help of the Pythagorean Theorem we can construct a figure composed of circular arcs that has an area equal to a triangle. You see, the inclusion of π in the circle-area formula usually causes a problem of equating circular areas to noncircular areas, which do not involve π. This is the result of the nature of π, an irrational number that can never be compared to rational numbers. Yet we will do just that here.

Let us consider a rather odd-shaped figure, a lune, which is a crescent-shaped figure (such as that in which the moon often appears) formed by two circular arcs.

The Pythagorean Theorem states that the sum of the areas of the squares on the legs of a right triangle is equal to the area of the square on the hypotenuse.

As a matter of fact, we can easily show that the "square" can be replaced by any similar figures (drawn appropriately) on the sides of a right triangle. The sum of the areas of the similar polygons on the legs of a right triangle is equal to the area of the similar polygon on the hypotenuse, such as are shown in figure 3-5.

This can then be restated for the specific case of semicircles (which are, of course, similar) to read: the sum of the areas of the semicircles on the legs of a right triangle is equal to the area of the semicircle on the hypotenuse.[5]

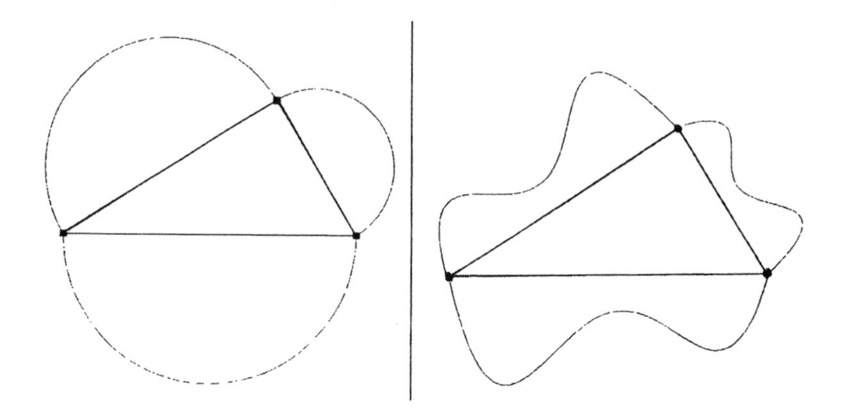

**Figure 3-5**

---

5. Since the area of a semicircle is $\frac{1}{2}\pi d^2$, where $d$ is its diameter, and for a right triangle $a^2 + b^2 = c^2$, then $\frac{1}{8}\pi a^2 + \frac{1}{8}\pi b^2 = \frac{1}{8}\pi c^2$.

Thus, for figure 3-6 we can say that the areas of the semicircles relate as follows:

$$\text{area } P = \text{area } Q + \text{area } R$$

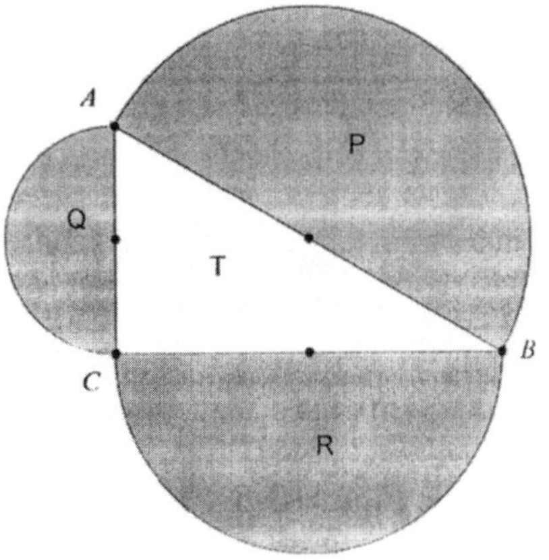

**Figure 3-6**

Suppose we now flip semicircle $P$ over the rest of the figure (using $AB$ as its axis) as shown in figure 3-7.

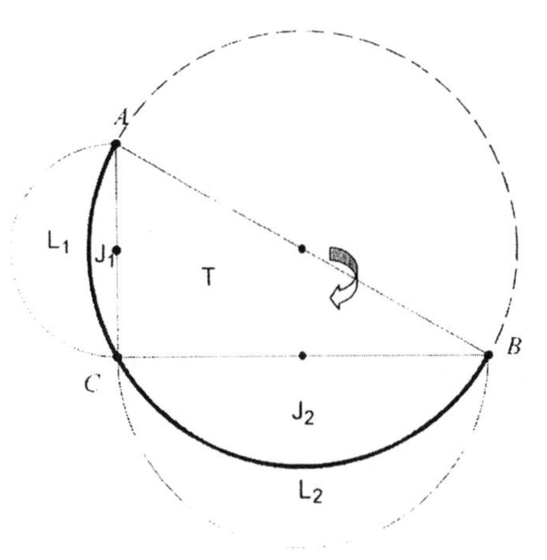


**Figure 3-7**

Let us now focus on the lunes formed by the two semicircles. We mark the lunes $L_1$ and $L_2$, as in figure 3-8.

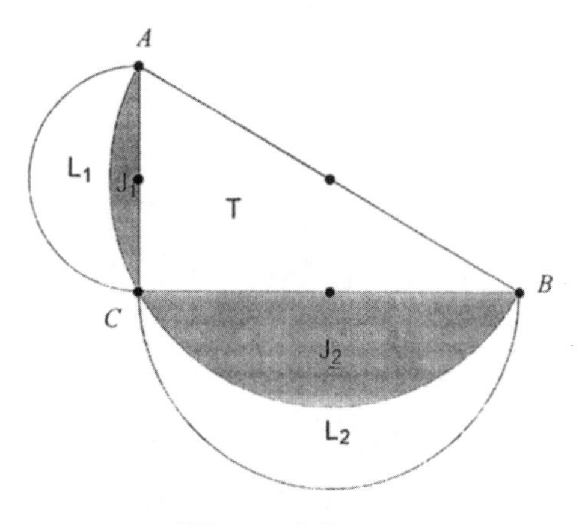


**Figure 3-8**

Earlier we established that area $P$ = area $Q$ + area $R$. In figure 3-8 that same relationship can be written as follows:

$$\text{area } J_1 + \text{area } J_2 + \text{area } T$$
$$= (\text{area } L_1 + \text{area } J_1) + (\text{area } L_2 + \text{area } J_2)$$

If we subtract $\text{area } J_1 + \text{area } J_2$ from both sides of the equation, we get the astonishing result:

$$\text{area } T = \text{area } L_1 + \text{area } L_2$$

That is, we have a rectilinear figure (the triangle) equal to some nonrectilinear figures (the lunes). This is quite unusual since the measures of circular figures seem to always involve $\pi$, while rectilinear (or straight line) figures do not.

## The Pythagorean Theorem Leads to Some Amazing Geometric Relationships

*When the Altitude Is Drawn to the Hypotenuse of a Right Triangle*

We begin with the familiar right triangle *ABC* with altitude *CD*, as shown in figure 3-9. We can show a very unusual extension of the Pythagorean Theorem here, namely, that $\frac{1}{h^2} = \frac{1}{a^2} + \frac{1}{b^2}$.

To show that this is a true statement, we will use a bit of algebra. We can write $\frac{1}{a^2} + \frac{1}{b^2} = \frac{a^2 + b^2}{a^2 b^2} = \frac{c^2}{a^2 b^2}$.

We can express the area of triangle $ABC$ in two ways: $\frac{ab}{2}$ and $\frac{hc}{2}$. Therefore, $ab = hc$ or $\frac{c}{ab} = \frac{1}{h}$. By squaring both sides of this equation we get $\frac{c^2}{a^2 b^2} = \frac{1}{h^2}$. Now replacing this in the previously obtained equation

$$\frac{1}{a^2} + \frac{1}{b^2} = \frac{c^2}{a^2 b^2}$$

gives us $\frac{1}{a^2} + \frac{1}{b^2} = \frac{1}{h^2}$, which is what we wanted to show.

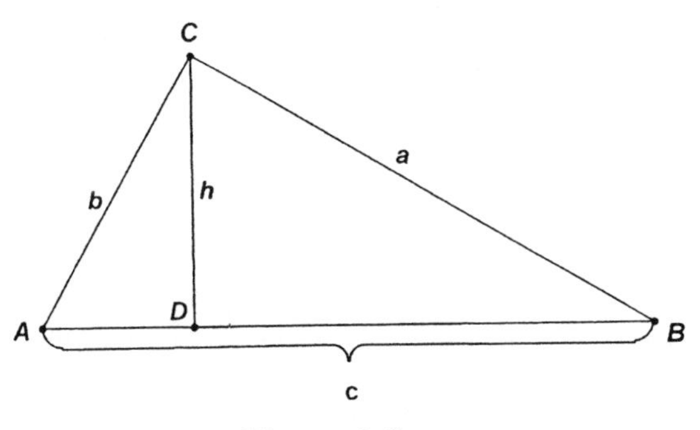


**Figure 3-9**

*A More General Triangle with Its Altitude*

Now we consider a randomly drawn triangle *ABC* (i.e., not necessarily a right triangle) with altitude *CD*, as shown in figure 3-10. By applying the Pythagorean Theorem to each of the triangles *ADC* and *BDC*, we have $h^2 = b^2 - p^2$ and $h^2 = a^2 - q^2$. Therefore, $b^2 - p^2 = a^2 - q^2$ and $b^2 - a^2 = p^2 - q^2$, which is true for any shaped triangle *ABC*. Notice the symmetry of this relationship.

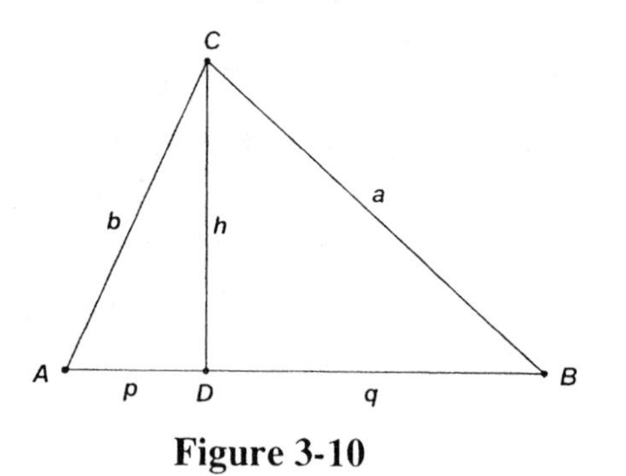


**Figure 3-10**

*The Isosceles Triangle with Its Altitude*

Now let's consider the *isosceles* triangle *ABC* with altitude *AD* as shown in figure 3-11.

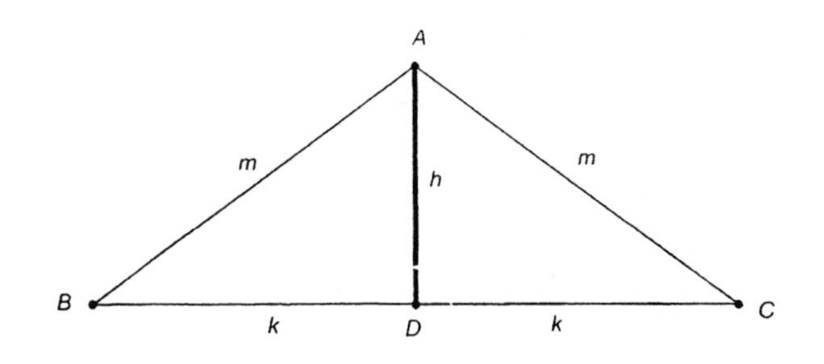


**Figure 3-11**

Since $\angle ADC = 90^\circ$, we can apply the Pythagorean Theorem to get $m^2 = h^2 + k^2$. This is simple enough. Now suppose we consider another segment from the vertex $A$ to another location, say, point $E$ on base $BC$, as shown in figure 3-12. We then get a very interesting and rather unexpected result. We can show that under these circumstances $m^2 = j^2 + pq$.

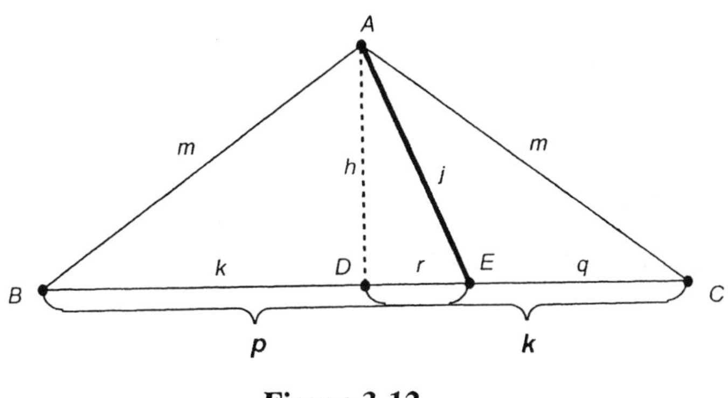


**Figure 3-12**

Using figure 3-12, we find that in comparing the two equations above that somehow $k^2$ was replaced by $pq$, when $h$ became length $j$. Before we see how this can be true (i.e., to prove it true), let's see what happens when the point $E$ moves farther along $BC$ toward $C$, to where it eventually coincides with point $C$. At that point, the length $q$ becomes 0, which results in the product $pq = 0$. Thus, we get the obvious (or what we would expect), namely, $m^2 = j^2 + 0$ and $m = j$.

Our task now is to justify the equation $m^2 = j^2 + pq$, which you will see is a direct outgrowth of the Pythagorean Theorem.

In right triangle $ADC$, $m^2 = AD^2 + DC^2$

In right triangle $ADE$, $j^2 = AD^2 + DE^2$

We will now subtract these two equations to get

$$m^2 - j^2 = DC^2 - DE^2$$

Factoring the right side of the equation gives us

$$m^2 - j^2 = (DC - DE)(DC + DE)$$

Since $BD = DC$,

$$m^2 - j^2 = (DC - DE)(BD + DE)$$
$$m^2 - j^2 = pq$$

Thus, we have shown that $m^2 = j^2 + pq$, which nicely relates any internal line segment from the vertex of an isosceles triangle to the side lengths.

*A Point Inside a Random Triangle*

The Pythagorean Theorem gives us a rather unusual relationship among the segments determined along the sides of the triangle by perpendiculars drawn from a point inside the triangle. What makes this particularly nice is that it is true for any triangle. As we see in figure 3-13, we select any point in a randomly constructed triangle. Here, we have point $P$ inside $\Delta ABC$. From point $P$, perpendiculars are drawn to meet the sides $BC$, $CA$, and $AB$ at points $D$, $E$, and $F$, respectively. It turns out that the following will always be true, regardless of the shape of the original triangle: The sum of the squares of the lengths of alternate segments on the sides of the triangle equals the sum of the squares of the lengths of the other three segments.

In figure 3-13, we have

$$BD^2 + CE^2 + AF^2 = DC^2 + EA^2 + FB^2$$

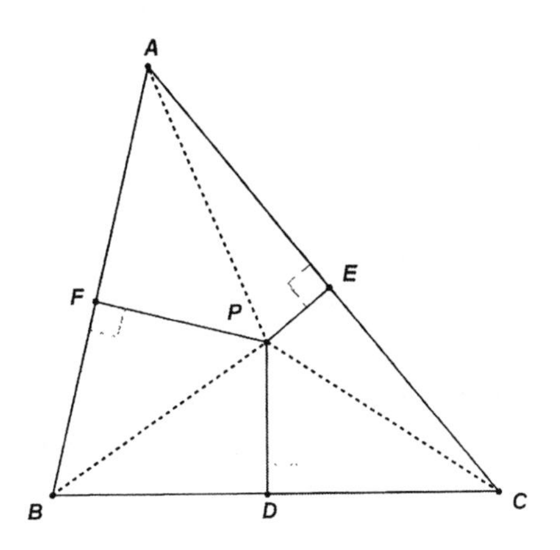


**Figure 3-13**

To see why this is true, we simply apply the Pythagorean Theorem to each of the six right triangles shown in figure 3-13:

For triangle $PDB$: $BD^2 + PD^2 = PB^2$
For triangle $PBF$: $FB^2 + PF^2 = PB^2$
Therefore, $BD^2 + PD^2 = FB^2 + PF^2$ (I)

For triangle $PDC$: $DC^2 + PD^2 = PC^2$
For triangle $PEC$: $CE^2 + PE^2 = PC^2$
Therefore, $DC^2 + PD^2 = CE^2 + PE^2$ (II)

For triangle $AEP$: $EA^2 + PE^2 = PA^2$
For triangle $AFP$: $AF^2 + PF^2 = PA^2$
Therefore, $EA^2 + PE^2 = AF^2 + PF^2$ (III)

By subtracting (II) from (I), we get

$$BD^2 - DC^2 = FB^2 + PF^2 - CE^2 - PE^2 \qquad \text{(IV)}$$

We can rewrite equation (III) in the form

$$EA^2 = AF^2 + PF^2 - PE^2$$

and subtract it from equation (IV) to obtain

$$BD^2 - DC^2 - EA^2 = FB^2 - CE^2 - AF^2 \quad \text{or}$$

$$BD^2 + CE^2 + AF^2 = DC^2 + EA^2 + FB^2$$

which is what we set out to demonstrate.

Again, what makes this so unusual is that it holds true for *any* triangle. You might want to see what happens when we repeat this for a point $P$ *outside* the triangle, as shown in figure 3-14.

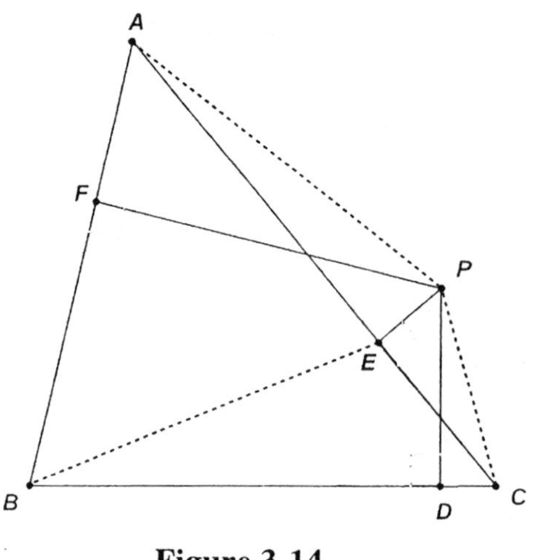


**Figure 3-14**

*A Random Point Inside a Rectangle*

This time we will randomly select a point in a rectangle and show a rather unexpected relationship among the distances from this randomly selected point to the vertices of the rectangle. In figure 3-15 we have a rectangle with a randomly selected point $P$. From this point we will draw the segments to each of the four vertices of the rectangle. For convenience we have marked the segment lengths.

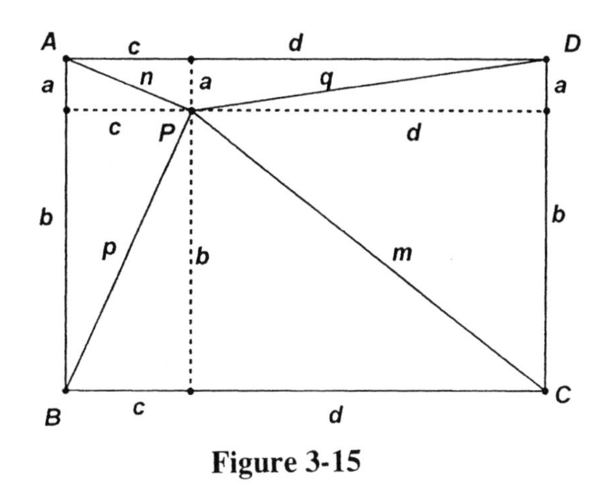


**Figure 3-15**

From the various right triangles shown in figure 3-15, we get the following Pythagorean Theorem relationships:

$$n^2 = a^2 + c^2$$
$$m^2 = b^2 + d^2$$
$$p^2 = b^2 + c^2$$
$$q^2 = a^2 + d^2$$

From these four equations we can see that

$$m^2 + n^2 = a^2 + b^2 + c^2 + d^2 = p^2 + q^2$$

What makes this so interesting is that this will be true for any point selected in the rectangle. Naturally, if the point $P$ is at the intersection of the diagonals of the rectangle, then the situation is certainly true—or, you might say, trivial. You may want to see if this will also hold true if the point $P$ is outside the rectangle.

*A Point on the Altitude of a Triangle*

When we consider the "random triangle"—one that could be of any shape—we can still find other surprising relationships, thanks to the Pythagorean Theorem. We will consider *any* $\Delta ABC$, where $E$ is any point on altitude $AD$. (See figure 3-16.) We can show that $AC^2 - CE^2 = AB^2 - EB^2$. Notice that when we say that point $E$ can be anywhere *on* altitude $AD$, it could also be at the endpoints, $A$ or $D$. If this is the case, then the relationship becomes trivial, since if $E$ coincides with $A$, then $AC = EC$ and $AB = EB$. While if $E$ coincides with $D$, then the Pythagorean Theorem tells us that $AC^2 - CE^2 = AD^2 = AB^2 - EB^2$.

We provide a simple proof for the points between $A$ and $D$ of this interesting relationship by simply applying the Pythagorean Theorem a few times.

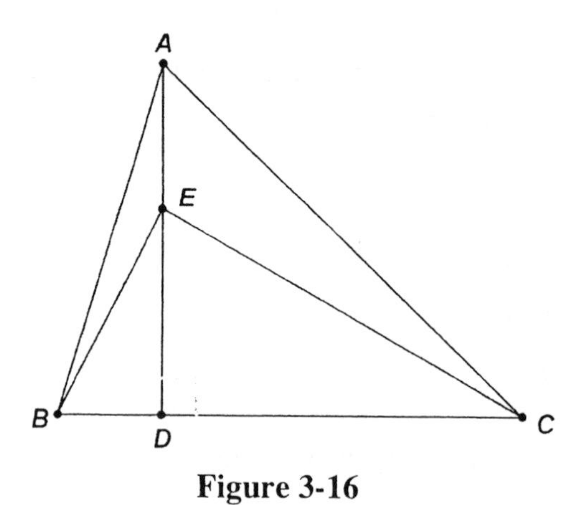


**Figure 3-16**

To prove that $AC^2 - CE^2 = AB^2 - EB^2$ for points between $A$ and $D$, we apply the Pythagorean Theorem as follows:

For $\Delta ADC$, $CD^2 + AD^2 = AC^2$
For $\Delta EDC$, $CD^2 + ED^2 = CE^2$

By subtraction, we get $AD^2 - ED^2 = AC^2 - CE^2$ (I)

Again applying the Pythagorean Theorem we get

For $\Delta ADB$, $DB^2 + AD^2 = AB^2$
For $\Delta EDB$, $DB^2 + ED^2 = EB^2$

Subtraction of these last two equations gives us

$$AD^2 - ED^2 = AB^2 - EB^2 \qquad \text{(II)}$$

Thus, from equations (I) and (II) we find our desired result: $AC^2 - CE^2 = AB^2 - EB^2$. Take note of the symmetry, which might best be described in words rather than merely symbolically. Suppose the angle at *B* were an obtuse angle. Then the altitude *AD* would lie outside the triangle. See if this relationship holds true for such an obtuse triangle.

*Points on the Legs of a Right Triangle*

Another nice relationship that the Pythagorean Theorem provides can be established if we take a randomly drawn right triangle and select two points on the legs and join them with a segment as shown in figure 3-17. The relationship is

$$BQ^2 + PC^2 = BC^2 + PQ^2$$

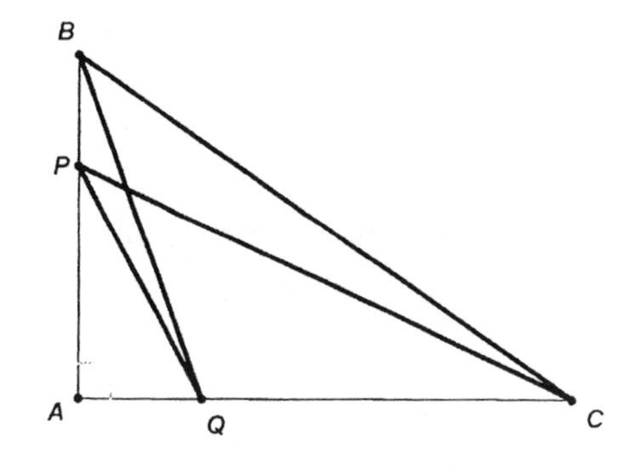


**Figure 3-17**

This is directly established from the Pythagorean Theorem.

Applying the Pythagorean Theorem to triangle $ABQ$:

$$BQ^2 = AQ^2 + AB^2$$

Applying the Pythagorean Theorem to triangle $APC$:

$$PC^2 = PA^2 + AC^2$$

Adding these two equations gives us

$$BQ^2 + PC^2 = \left(AQ^2 + PA^2\right) + \left(AB^2 + AC^2\right) \qquad (*)$$

Applying the Pythagorean Theorem again, but this time to triangles $PAQ$ and $ABC$:

$$AQ^2 + PA^2 = PQ^2 \text{ and } AB^2 + AC^2 = BC^2$$

Then substituting these two equations into the previous one we arrive at the desired result:

$$BQ^2 + PC^2 = PQ^2 + BC^2$$

Take one more look at this elegant relationship and treasure it. Generalize it in words.

Other interesting relationships can be crafted from this figure. If we go back to equation (*) and replace only $AB^2 + AC^2 = BC^2$, we get

$$BQ^2 + PC^2 = \left(AQ^2 + PA^2\right) + BC^2$$
$$BQ^2 - AQ^2 + PC^2 - PA^2 = BC^2$$

This is also a rather nice relationship solely dependent on the Pythagorean Theorem.

*The Median and the Right Triangle*

The median[6] drawn to the hypotenuse of a right triangle has a very special property: it is half the length of the hypotenuse. (A proof of this useful relationship may be found in appendix A.) With this relationship and the Pythagorean Theorem we can establish another interesting triangle relationship: namely, the sum of the squares of the sides of a right triangle is equal to eight times the square of the median drawn to the hypotenuse. In figure 3-18, we will show that $AC^2 + BC^2 + AB^2 = 8(CD^2)$.

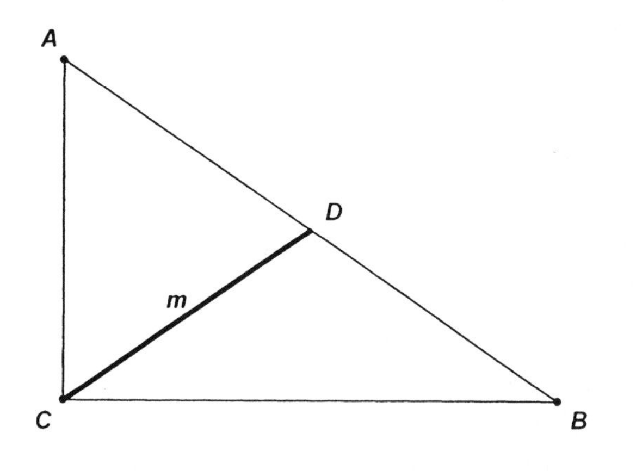


**Figure 3-18**

This is rather easy to demonstrate—again with the simple application of the Pythagorean Theorem applied to right triangle $ABC$: $AC^2 + BC^2 = AB^2$. From the above-stated relationship between the hypotenuse and the median drawn to it, we have $AB = 2CD$. Therefore, $AB^2 = 4(CD^2)$. The conclusion is then easily obtained by adding the equations and doing a substitution.

6. The median of a triangle is the segment joining a vertex and the midpoint of the opposite side.

$$AC^2 + BC^2 = AB^2$$
$$AC^2 + BC^2 + AB^2 = AB^2 + 4\left(CD^2\right)$$
$$AC^2 + BC^2 + AB^2 = 4\left(CD^2\right) + 4\left(CD^2\right)$$
$$AC^2 + BC^2 + AB^2 = 8\left(CD^2\right)$$

You might find another variation of this relationship interesting, that is, that the sum of the squares of the legs of a right triangle is equal to four times the square of the median to the hypotenuse, or $AC^2 + BC^2 = 4\left(CD^2\right)$, which is arrived at by replacing the $AB^2 = 4\left(CD^2\right)$—from the equation above—in the Pythagorean Theorem: $AC^2 + BC^2 = AB^2$.

*Medians to the Legs of a Right Triangle*

Another nifty relationship that be can established directly from the Pythagorean Theorem is that for any right triangle, four times the sum of the squares of the medians drawn to the legs of the triangle is equal to five times the square of the triangle's hypotenuse. In figure 3-19—with median lengths $m$ and $n$—what we're claiming is that $4\left(m^2 + n^2\right) = 5c^2$, where $c$ is the length of the hypotenuse of triangle $ABC$.

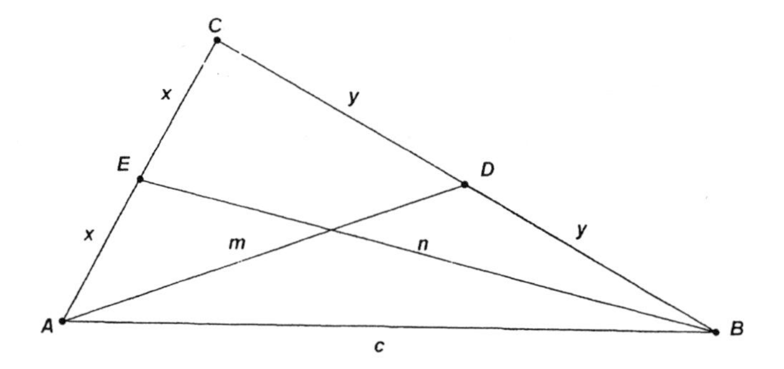


**Figure 3-19**

Here we would apply the Pythagorean Theorem to triangles *ADC*, *BEC*, and *ABC*. Beginning with $\Delta ADC: m^2 = y^2 + (2x)^2$.

Then for triangle *BEC*: $n^2 = x^2 + (2y)^2$.

By adding these two equations we get

$$m^2 + n^2 = x^2 + y^2 + 4(x^2 + y^2) = 5(x^2 + y^2) \qquad \text{(I)}$$

However, by applying the Pythagorean Theorem to triangle *ABC*, we get

$$c^2 = (2x)^2 + (2y)^2 = 4(x^2 + y^2)$$

This can be rewritten as $\frac{c^2}{4} = x^2 + y^2$ (II)

By substituting equation (II) into equation (I) we find that

$$m^2 + n^2 = 5\left(\frac{c^2}{4}\right) = \frac{5}{4}c^2$$

Multiplying both sides by 4, we get the relationship we set out to demonstrate: $4(m^2 + n^2) = 5c^2$.

*The Relationship of the Medians and the Sides of a Random Triangle*

There are many wonderful relationships that emerge when we apply the Pythagorean Theorem. One quite amazing relationship is that $\frac{3}{4}$ of the sum of the squares of the sides of any triangle is equal to the sum of the squares of the medians of the triangle. In other words, $\frac{3}{4}(a^2+b^2+c^2)=m_a^{\ 2}+m_b^{\ 2}+m_c^{\ 2}$ (see figure 3-20).[7]

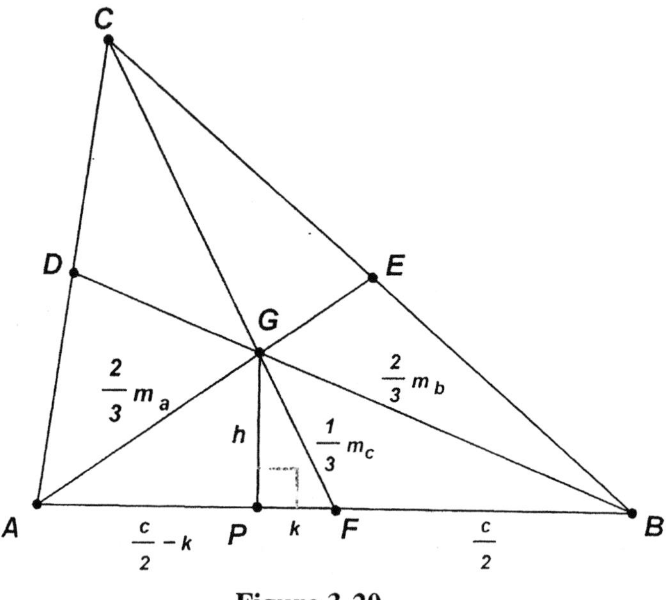


**Figure 3-20**

To prove this true for all triangles, we will begin with the "randomly drawn" $\Delta ABC$ with medians[8] $AE$, $BD$, and $CF$. We will then draw $GP \perp AB$ ($\perp$ means "is perpendicular to") as shown in figure 3-20. For convenience we will let $GP = h$ and $PF = k$. Since $F$ is the midpoint of $AB$, we have $AF = \frac{c}{2}$, and then $AP = \frac{c}{2} - k$.

Now we will use the fact that $AG = \frac{2}{3}AE$ and then apply the Pythagorean Theorem to $\Delta AGP$.

---

7. We usually label the sides opposite the angles $A$, $B$, and $C$ as $a$, $b$, and $c$, respectively. We will use $m_c$ to refer to the measure of the median drawn to the side $c$.

8. It will be helpful to recall that the medians of a triangle trisect each other and meet at a common point, called the *centroid*, which is the center of gravity of the triangle.

$$h^2 + \left(\frac{c}{2} - k\right)^2 = \left(\frac{2}{3}m_a\right)^2$$

This can then be rewritten as $h^2 + \frac{c^2}{4} - ck + k^2 = \frac{4}{9}m_a^{\ 2}$ (I)

Applying the Pythagorean Theorem to $\Delta BGP$,

$$h^2 + \left(\frac{c}{2} + k\right)^2 = \left(\frac{2}{3}m_b\right)^2$$

which reduces to $h^2 + \dfrac{c^2}{4} + ck + k^2 = \dfrac{4}{9}m_b^{\ 2}$ (II)

Adding (I) and (II) gives us

$$2h^2 + \frac{2c^2}{4} + 2k^2 = \frac{4}{9}m_a^{\ 2} + \frac{4}{9}m_b^{\ 2}$$

or $2h^2 + 2k^2 = \dfrac{4}{9}m_a^{\ 2} + \dfrac{4}{9}m_b^{\ 2} - \dfrac{c^2}{2}$ (III)

However, in $\Delta FGP$, the Pythagorean Theorem gives us

$$h^2 + k^2 = \left(\frac{1}{3}m_c\right)^2$$

Therefore, $2h^2 + 2k^2 = \frac{2}{9}m_c^{\ 2}$ (IV)

By substitution of (IV) into (III),

$$\frac{2}{9}m_c^{\ 2} = \frac{4}{9}m_a^{\ 2} + \frac{4}{9}m_b^{\ 2} - \frac{c^2}{2}$$

Therefore, $c^2 = \dfrac{8}{9}m_a^{\ 2} + \dfrac{8}{9}m_b^{\ 2} - \dfrac{4}{9}m_c^{\ 2}$

Similarly, $b^2 = \frac{8}{9}m_a^{\ 2} + \frac{8}{9}m_c^{\ 2} - \frac{4}{9}m_b^{\ 2}$

And analogously, $a^2 = \frac{8}{9}m_b^{\ 2} + \frac{8}{9}m_c^{\ 2} - \frac{4}{9}m_a^{\ 2}$

By adding the last three equations,

$$a^2 + b^2 + c^2 = \frac{4}{3}(m_a^{\ 2} + m_b^{\ 2} + m_c^{\ 2})$$

$$\text{or} \quad \frac{3}{4}(a^2 + b^2 + c^2) = m_a^{\ 2} + m_b^{\ 2} + m_c^{\ 2}$$

This is what we set out to establish.

*More about the Medians of a Right Triangle*

If in figure 3-21, $AE$ and $BF$ are medians drawn to the legs of right $\Delta ABC$, there is an interesting value for $\frac{(AE)^2+(BF)^2}{(AB)^2}$. We will determine this value here with the help of the Pythagorean Theorem.

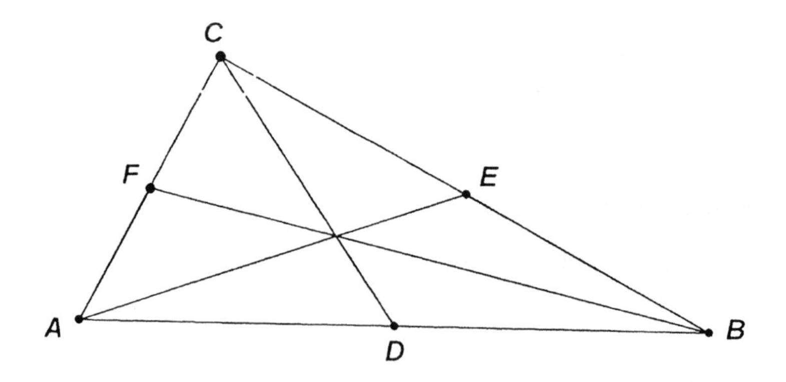


**Figure 3-21**

We just established (above) that the sum of the squares of the measures of the medians equals $\frac{3}{4}$ the sum of the squares of the measures of the sides of the triangle.

$$AE^2 + BF^2 + CD^2 = \frac{3}{4}\left(AC^2 + CB^2 + AB^2\right) \quad \text{(I)}$$

By the Pythagorean Theorem: $AC^2 + CB^2 = AB^2$ (II)

Also, $CD = \frac{1}{2}(AB)$ (III)

By substituting equations (II) and (III) into (I), we get

$$AE^2 + BF^2 + \left(\frac{1}{2}AB\right)^2 = \frac{3}{4}\left(AB^2 + AB^2\right)$$

or $AE^2 + BF^2 = \frac{3}{2}AB^2 - \frac{1}{4}AB^2 = \frac{5}{4}AB^2$

Then $\frac{AE^2 + BF^2}{AB^2} = \frac{5}{4}$

This is a rather surprising result since we began this without any unit measure established. What it tells us is that the ratio of the sum of the squares of the two medians $AE^2 + BF^2$ to the square of the hypotenuse $AB^2$ is 5 to 4. This can be visually shown in figure 3-22, where the sum of the areas of the two squares (*AELK* and *BFGH*) on the medians (drawn to the legs) is equal to $\frac{5}{4}$ the area of the square (*ABNM*) on the hypotenuse.

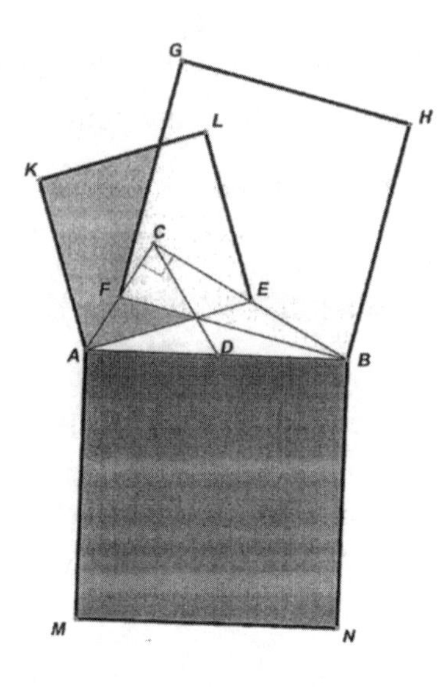


**Figure 3-22**

*Surprises on the Classic Figure of the Pythagorean Theorem*

Recall the classic illustration of the Pythagorean Theorem, shown in figure 3-23.

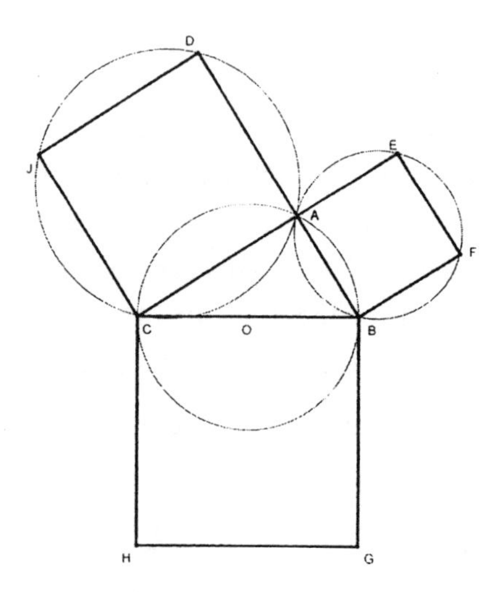


**Figure 3-23**

Consider the circumscribed circles[9] of the squares *ADJC* and *BFEA*. They are tangent at point *A*, which is on the circumcircle of the right triangle *ABC*. Putting this another way, we can conclude that the two circumcircles of the squares drawn on the legs of a right triangle have a point in common with the circle whose diameter is the third side (namely, the hypotenuse). This is not too amazing, but if we change this to a more general case, where we no longer consider a right triangle but rather any randomly drawn triangle, then we have something quite spectacular.

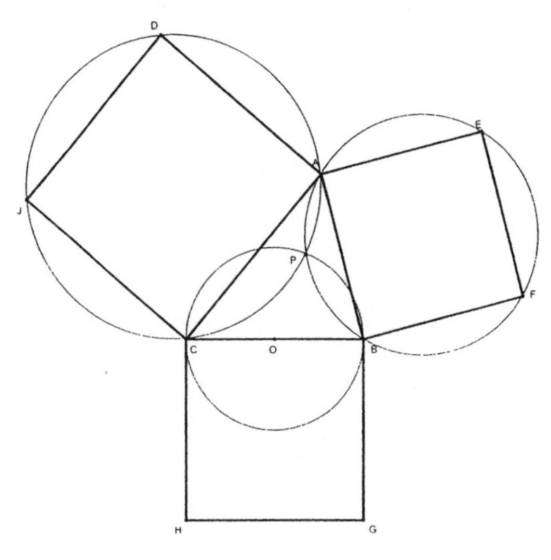


**Figure 3-24**

Rather than having the right triangle as in figure 3-23, we shall begin with a randomly drawn triangle, $\Delta ABC$, and then draw squares on sides *AB* and *AC* as shown in figure 3-24. The circumcircles of these two squares meet at point *P*. When we construct the circle with *BC* as its diameter, we find that it also contains the point *P*. The location of point *P*—being inside $\Delta ABC$ or outside it—depends on the measure of $\angle BAC$. What might be expected

9. The *circumscribed circle* (or *circumcircle*) of a polygon is a circle where all vertices of the polygon lie on the circle.

regarding its location? Needless to say, this is quite astonishing, and, at the same time, not easy to discover.

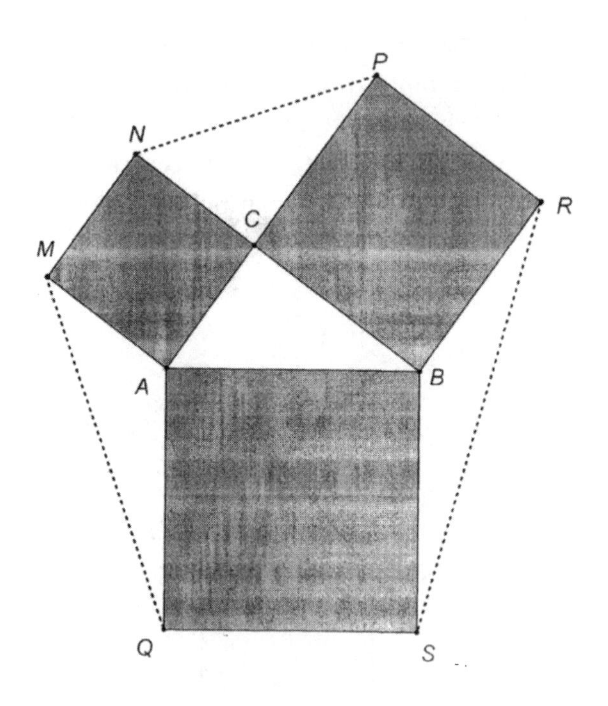


**Figure 3-25**

While we are inspecting the now-famous illustration of the Pythagorean Theorem shown in figure 3-25, we add to this diagram the line segments *MQ*, *NP*, and *RS*, as shown on figure 3-26. We could show a nice extension of the Pythagorean Theorem: that the sum of the squares of these three segments equals three times the sum of the squares of the sides of the original right triangle.[10] That is,

$$MQ^2 + NP^2 + RS^2 = 3\left(AB^2 + BC^2 + AC^2\right)$$

(See appendix A for a proof of this relationship.)

10. This doesn't have to be a right triangle. It will be true for any $\Delta ABC$ (see appendix A).

Now let's consider the triangles formed between the squares as shown in figure 3-27. They are $\Delta AMQ$, $\Delta BRS$, and $\Delta NCP$.

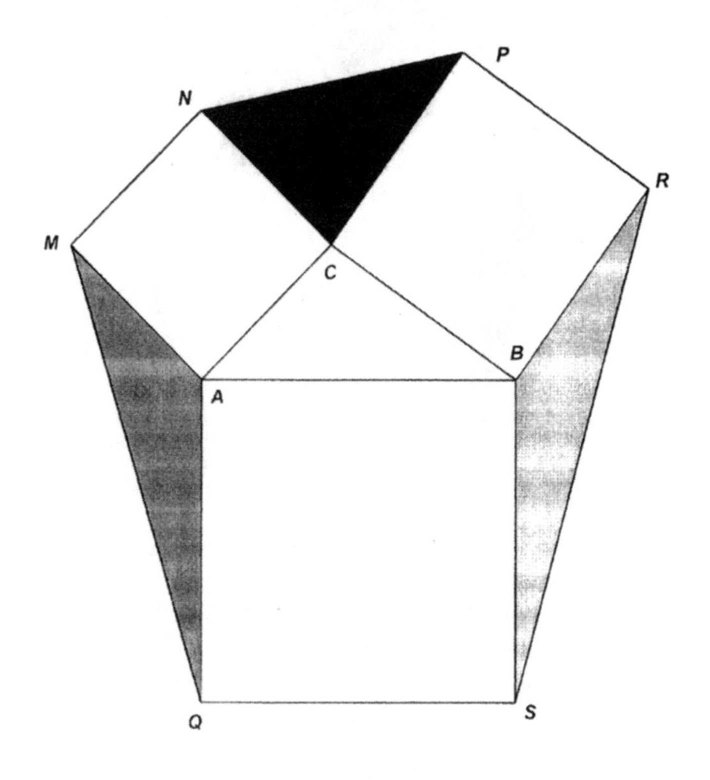


**Figure 3-26**

We shall isolate these three triangles and then show graphically that they are not only equal in area but also equal to the area of the original right triangle.[11] (See figure 3-27.)

---

11. For those who would like to see this done directly (without graphic manipulations), recall the formula for the area of a triangle is $\frac{1}{2}ab\sin C$.

Applied to $\Delta MAQ$, the area $= \frac{1}{2}(MA)(AQ)\sin\angle MAQ$.

The area of $\Delta ABC = \frac{1}{2}(AC)(AB)\sin\angle BAC$.

Since $\angle BAC = 180° - \angle MAQ$, $\sin\angle MAQ = \sin\angle BAC$.

Also $MA = AC$ and $AQ = AB$.

Therefore, the triangles $MAQ$ and $ABC$ have equal areas.

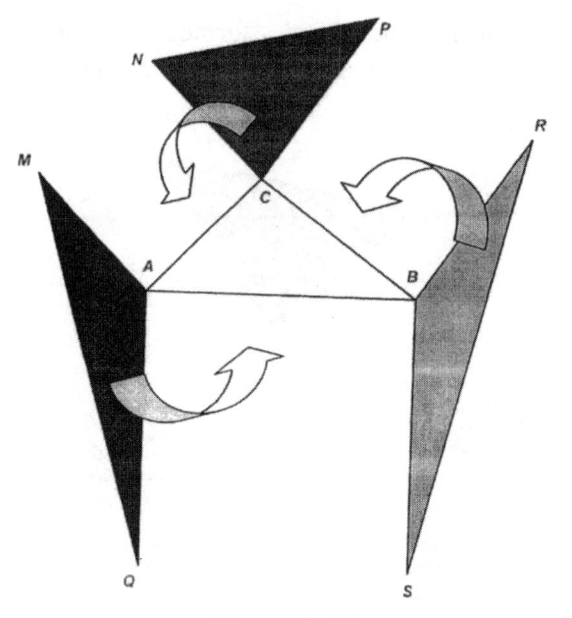


**Figure 3-27**

As we show in figure 3-27, we now rotate triangle *NPC* on point *C* until *NC* coincides with *AC*.

Then we rotate triangle *MQA* on point *A* until *AQ* coincides with *AB*.

Finally, we rotate triangle *SRB* on point *B* until *RB* coincides with *CB*.

The result is shown in figure 3-28.

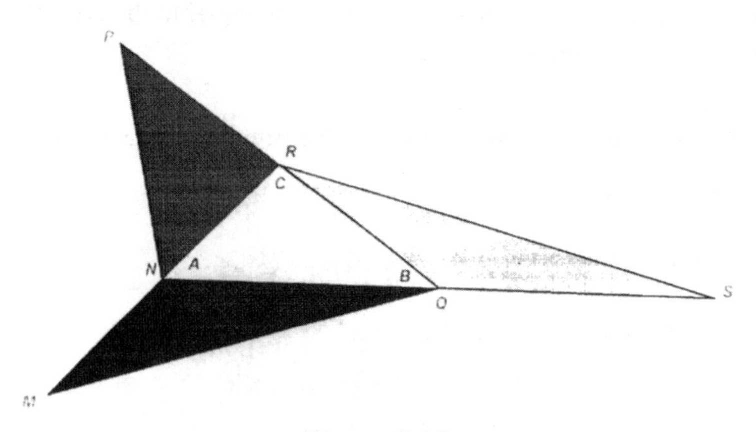


**Figure 3-28**

It is natural to wonder why the four triangles all have the same area. This can be easily deduced from figure 3-28. Since the bases of the two triangles *NPC* and *ABC* are equal (*PC* = *BC*), and *PCB* is a straight line (two right angles meet at point *C*), and the altitude from point *N* (or *A*) to the line *PCB* is common to both triangles *NPC* and *ABC*, they must have equal areas. The same reasoning is then used to show that triangle *RSB* is equal in area to triangle *ABC*—since *AB* = *BS* and the altitude from *C* (or *R*) is shared by both triangles. Similarly, triangle *MQA* has the same area as triangle *ABC*. Therefore, the four triangles have the same area.

### *The Famous "Pythagorean Figure" Yields Another Property*

The famous "Pythagorean figure" to which we refer here is that shown in figure 3-29.

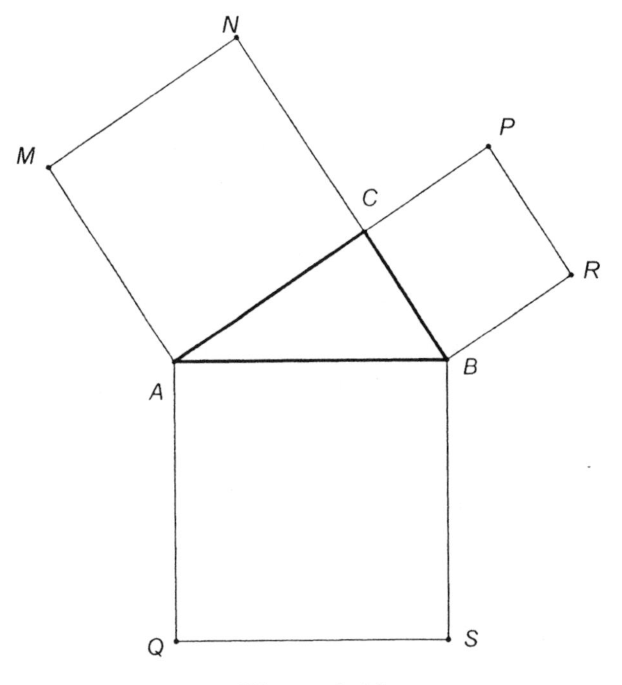


**Figure 3-29**

This time we will simply focus on the centers of the three squares drawn on the sides of a right triangle as shown in figure 3-30.

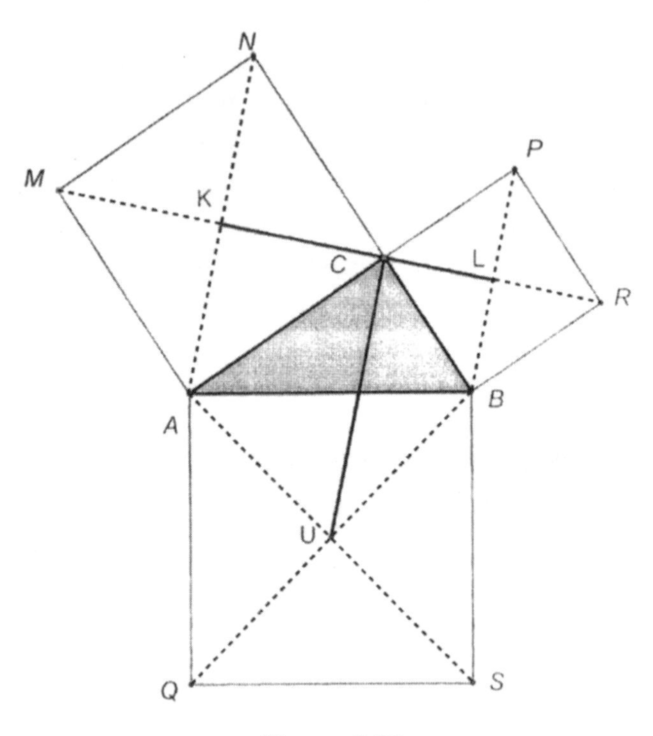


**Figure 3-30**

We claim that the two interior line segments joining the centers are equal in length and perpendicular to each other. To prove this to be true, we will revert to the figure (figure 2-15) we used in chapter 2 (Demonstration 7) to prove the Pythagorean Theorem. We provide that figure again here (figure 3-31) with additional line segments drawn and focus on two quadrilaterals that will be the key to our demonstration.

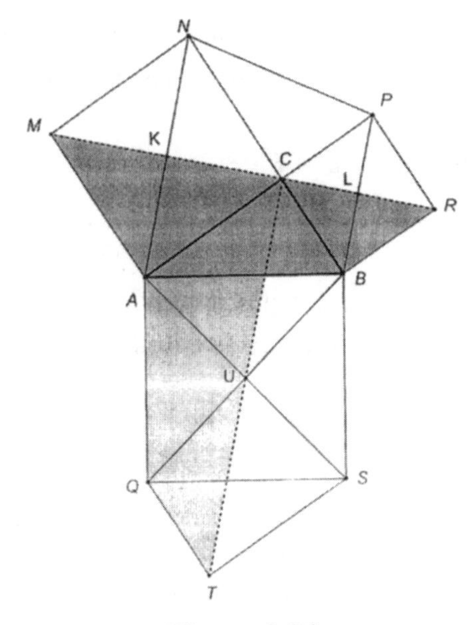


**Figure 3-31**

We can show that the two shaded quadrilaterals (*MABR* and *CAQT*) are congruent. Therefore, $\angle ACT = \angle AMR = 45°$. We also know that $\angle ACM = 45°$; thus $\angle MCT = 90°$, which concludes the perpendicularity part of this demonstration. Going back to the congruent quadrilaterals, we can see that $MR = CT$. Since *KC* and *CL* are each half of the two segments that compose *MR*, we can conclude that $KL = \frac{1}{2}MR$. By an apparent symmetry, we can also conclude that $CU = \frac{1}{2}CT$. Therefore, *KL* = *CU*, which is what we set out to demonstrate. This is a rather surprising result, given that it is true for all shapes of right triangles!

*The Pythagorean Theorem on the Rectangle*

Another extension of the Pythagorean Theorem can be found in the rectangle shown in figure 3-32. For rectangle *ABCD*, $m^2 + n^2 = a^2 + b^2 + c^2 + d^2$. This says that the sum of the squares of the diagonals of a rectangle equals the sum of the squares of the sides.

We can justify (i.e., prove) this statement by applying the Pythagorean Theorem to each of the two triangles *ABC* and *ABD*.

For $\Delta ABD: \ n^2 = c^2 + b^2$

For $\Delta ABC: m^2 = c^2 + d^2$

Therefore, by realizing that $c = a$ and adding these two equations we immediately get our desired result:

$$m^2 + n^2 = a^2 + b^2 + c^2 + d^2$$

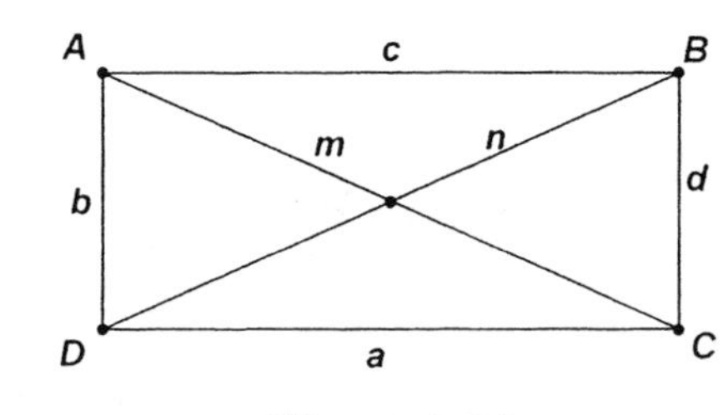


**Figure 3-32**

If we stretch this rectangle to form a parallelogram, this relationship, much to our amazement, still holds true, even though we are not applying the Pythagorean Theorem.

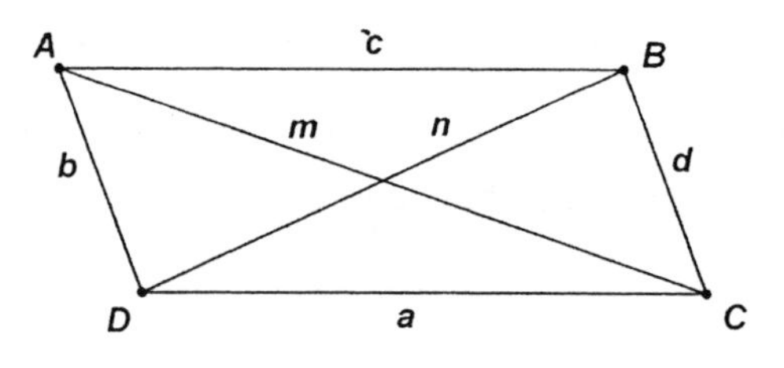


**Figure 3-33**

That is, for the parallelogram in figure 3-33, the relationship $m^2 + n^2 = a^2 + b^2 + c^2 + d^2$ remains true. (See appendix A for a proof.)

*The Pythagorean Theorem Provides an Alternative Formula for the Area of a Right Triangle*

Consider the right triangle with the circle inscribed in it as shown in figure 3-34. We will show through some simple algebra and the Pythagorean Theorem that the area is simply the product of the two segments on the hypotenuse; that is, the area of right triangle $ABC$ is equal to $pq$.

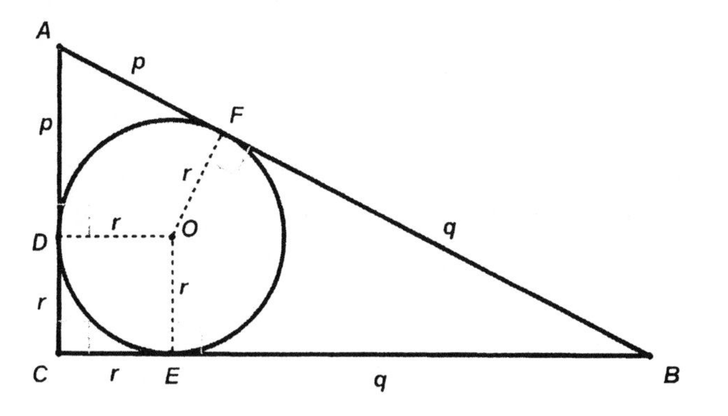


**Figure 3-34**

The area of right triangle $ABC$ equals

$$\frac{AC \cdot BC}{2} = \frac{(p+r)(q+r)}{2} = \frac{pq + pr + qr + r^2}{2}$$

However, by the Pythagorean Theorem, we get

$$(p+r)^2 + (q+r)^2 = (p+q)^2$$

which with simple algebra reduces to

$$pr + qr + r^2 = pq$$

Thus, the last three terms in the numerator of our previous expression for the area of the right triangle add up to $pq$. Therefore,

$$\text{area of right triangle } ABC = \frac{2pq}{2} = pq$$

a rather simple way to express the area of the triangle.

## The Pythagorean Theorem in Three Dimensions

*Linear Extension*

We shall now extend the Pythagorean Theorem to three dimensions in the form of a rectangular solid[12] to get an interesting analog: $a^2 + b^2 + c^2 = d^2$. (See figure 3-35.)

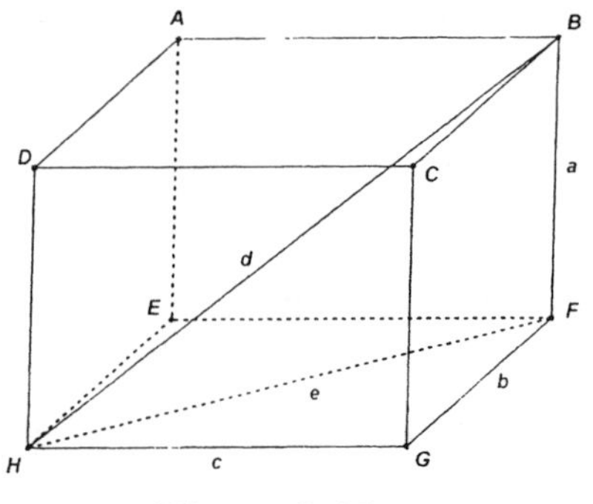


**Figure 3-35**

In figure 3-35 we show a rectangular solid, where we can apply the Pythagorean Theorem twice.

For right triangle $HGF$: $b^2 + c^2 = e^2$
For right triangle $BFH$: $a^2 + e^2 = d^2$
By substituting for $e^2$ in the second equation above, we get

$$a^2 + b^2 + c^2 = d^2$$

12. This figure can also be referred to as a *rectangular parallelepiped* or a *cuboid*.

*Area Analog of the Pythagorean Theorem*

Working further with the rectangular solid we can compare areas of various parts—each time you will see that not only will we need the Pythagorean Theorem, but you will notice an interesting analog for it.

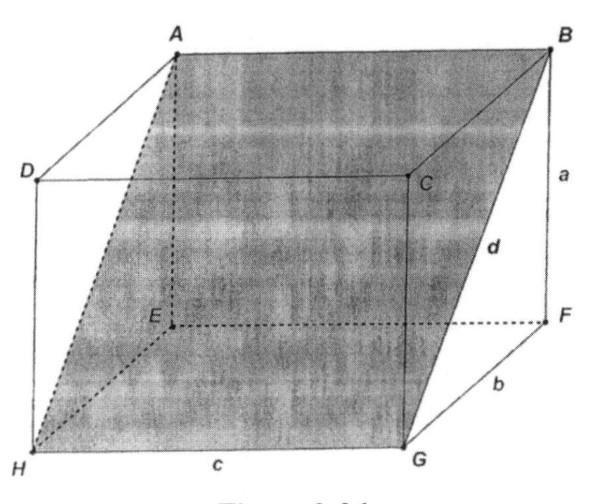


**Figure 3-36**

Consider the rectangular solid shown in figure 3-36. The square of the area of rectangle *ABGH* (shaded) is equal to the sum of the squares of the areas of rectangle *HEFG* (bottom) and rectangle *ABFE* (rear). This is rather simple to demonstrate. The area of rectangle *ABGH* = *cd*. However, by applying the Pythagorean Theorem, $d = \sqrt{a^2 + b^2}$. So the area of rectangle *ABGH* = $c\sqrt{a^2 + b^2}$.

Therefore, $(\text{area rect. } ABGH)^2 = c^2(a^2 + b^2) = c^2a^2 + c^2b^2$

Thus we have

$$(\text{area rect. } ABGH)^2 = (\text{area rect. } ABFE)^2 + (\text{area rect. } HEFG)^2$$

Notice the three-dimensional analogs of the Pythagorean Theorem for the prism.

*More Analogs on the Rectangular Solid*

We will now compare areas of triangles on this rectangular solid. Actually we will cut off a corner and consider this tetrahedron $ACHD$. (See figure 3-37.)

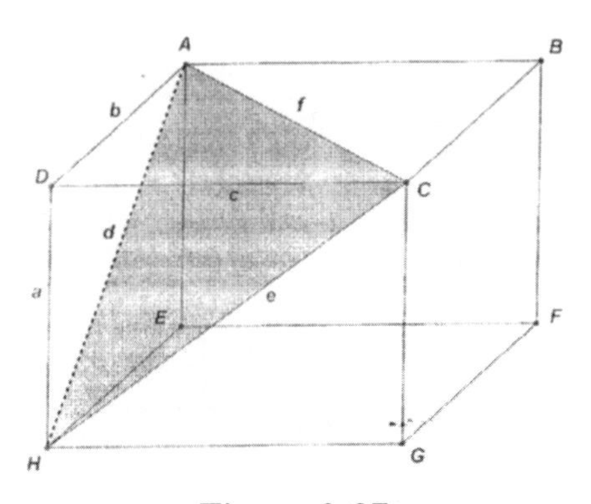


**Figure 3-37**

We shall isolate the tetrahedron $ACHD$ (i.e., the corner solid of four faces that we cut from the rectangular solid in figure 3-37), shown in figure 3-38.

We will now take you through a brief algebraic journey to an amazing result: namely, that the sum of the squares of the areas of the three right triangle faces of the tetrahedron equals the square of the fourth side of the tetrahedron.

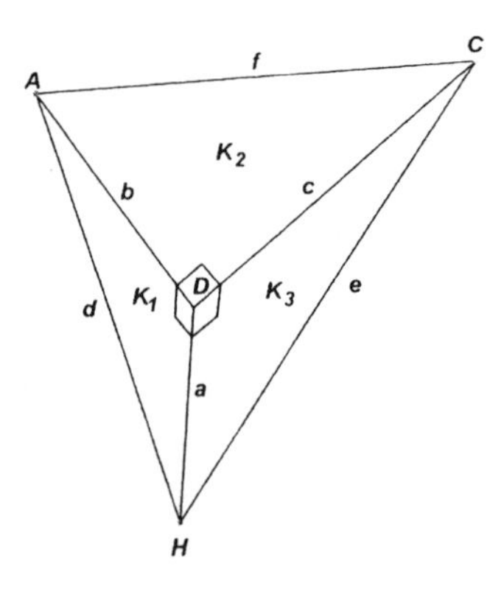


**Figure 3-38**

Symbolically (figure 3-38), if we let

$$\text{area } \Delta ADH = K_1$$
$$\text{area } \Delta ADC = K_2$$
$$\text{area } \Delta CDH = K_3$$
$$\text{area } \Delta ACH = K$$

we would show that $K_1^2 + K_2^2 + K_3^2 = K^2$. We could actually refer to the areas of the right triangle faces $(K_1, K_2, K_3)$ as taking the role of the "legs" and the area of the third face of triangle ($K$) taking the role as the hypotenuse of the tetrahedron.

A small algebraic excursion—some simple elementary algebra—would be in order here. We begin by noting that

$$K_1 = \frac{1}{2}ab$$
$$K_2 = \frac{1}{2}bc$$
$$K_3 = \frac{1}{2}ac$$

To determine the area of the fourth face of the tetrahedron we will use Heron's formula (see page 67).

When we apply this formula to triangle $ACH$, the following results:

$$K = \sqrt{s(s-d)(s-e)(s-f)}$$

Recall, $s$, the semiperimeter, is $\frac{1}{2}(d+e+f)$.

Squaring both sides of the last equation and replacing $s$, we get

$$K^2 = \frac{1}{16}(d+e+f)(-d+e+f)(d-e+f)(d+e-f)$$

With some algebraic manipulation we can rewrite this as

$$K^2 = \frac{1}{16}\left[\left(d^2+e^2+f^2\right)^2 - 2\left(d^4+e^4+f^4\right)\right]$$

When we apply the Pythagorean Theorem to the three right triangular faces we get

$$d^2 = a^2 + b^2$$
$$e^2 = a^2 + c^2$$
$$f^2 = b^2 + c^2$$

By adding these, we get

$$d^2 + e^2 + f^2 = 2\left(a^2 + b^2 + c^2\right)$$

When we add the squares of each of the Pythagorean Theorem equations above we get

$$d^4 + e^4 + f^4 = 2\left(a^4 + b^4 + c^4\right) + 2\left(a^2b^2 + b^2c^2 + a^2c^2\right)$$

With appropriate substitution into the alternate form of Heron's formula

$$K^2 = \frac{1}{16}\left[\left(d^2+e^2+f^2\right)^2 - 2\left(d^4+e^4+f^4\right)\right]$$

we get the complicated-looking expression (fear not, it will be simplified):

$$K^2 = \frac{1}{16}\left[\left(2\left(a^2+b^2+c^2\right)\right)^2 - 2\left[2\left(a^4+b^4+c^4\right) + 2\left(a^2b^2+b^2c^2+a^2c^2\right)\right]\right]$$
$$K^2 = \frac{1}{4}\left[a^2b^2 + b^2c^2 + a^2c^2\right]$$

We are almost at the conclusion of this algebraic "exercise"—we just need to replace the terms in the above equations with their area representations

$$\left(K_1 = \frac{1}{2}ab,\ K_2 = \frac{1}{2}bc, \text{ and } K_3 = \frac{1}{2}ac\right)$$

that we established at the beginning of this development. We obtain $K^2 = K_1^2 + K_2^2 + K_3^2$, which is what we set out to show.

As an extra bonus for this tetrahedron, we can make an additional statement. The square of the volume of the tetrahedron is equal to $\frac{2}{9}$ of the product of the areas of the three right triangle faces. Symbolically: $V^2 = \frac{2}{9}K_1K_2K_3$.

The enthusiast will have many more analogues to discover in this three-dimensional figure.[13]

*A Rectangular Solid with Integer Edges and Diagonals?*

The Pythagorean Theorem has already been extended to the rectangular solid in that $a^2 + b^2 + c^2 = g^2$. The question then is: Can a rectangular solid be constructed where the three edges (length, width, and height) have integer lengths and the distance between any two vertices is also an integer length? (See figure 3-39.)

13. See A. S. Posamentier and L. R. Patton, "Enhancing Plane Euclidean Geometry with Three-Dimensional Analogs," *Mathematics Teacher* 102, no. 5 (December 2008/January 2009): 394–98.

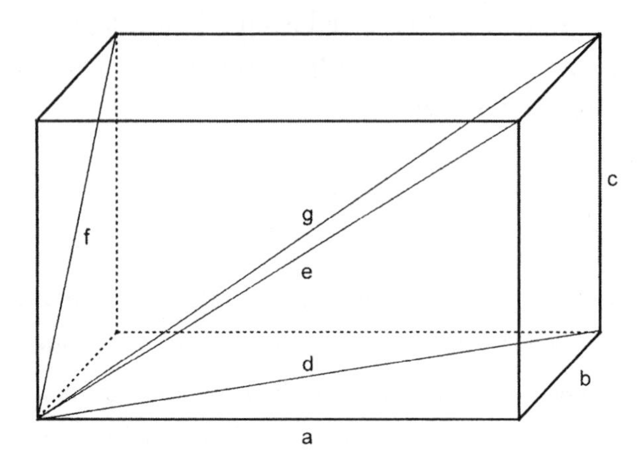


**Figure 3-39**

To investigate this, we must be able to apply the Pythagorean Theorem to show that each of the following equations can be solved with only integers:

$$a^2 + b^2 = d^2$$
$$a^2 + c^2 = e^2$$
$$b^2 + c^2 = f^2$$
$$a^2 + b^2 + c^2 = g^2$$

The prolific Swiss mathematician Leonhard Euler (1707–1783) almost found a solution to this problem with the following values: $a = 240$, $b = 44$, $c = 117$, $d = 244$, $e = 267$, and $f = 125$. However, his effort fell short of the complete solution to the problem, since the last of these equations ($a^2 + b^2 + c^2 = g^2$) does not result in an integer solution; $g \approx 270.60$. It was not until the year 2000 that Marcel Lüthi, a Swiss mathematics teacher, proved that such a rectangular solid with integer distances between all vertices could not exist.[14] As often happens

14. Marcel Lüthi, "Zum Problem rationaler Quader," *Praxis der Mathematik* 42 (2000): 177.

in mathematics, we have a situation where a problem remained unsolved for several hundred years and only recently has been solved.

## Pythagorean Magic Squares

A magic square is a square matrix of numbers in which the sum of the numbers in each of its columns, rows, and diagonals is the same. There are entire books written about magic squares of all kinds.[15] There is one magic square, however, that stands out from among the rest for its beauty and extended properties. This magic square even comes to us through art, *Melencolia I* (figure 3-40), rather than merely through the usual mathematical channels. It is depicted in the background of the famous engraving produced in 1514 by renowned German artist Albrecht Dürer (1471–1528).

Most of Dürer's works were signed by him with his initials, one over the other, and with the year in which the work was made. Here we find it in the shaded region near the lower right side of the picture. We notice that it was etched in the year 1514. The observant reader may notice that the two center cells of the bottom row depict the year as well. Let us look at this magic square more closely (figure 3-41).

---

15. Two books to be recommended are W. H. Benson and O. Jacovy, *New Recreations with Magic Squares* (New York: Dover, 1976) and W. S. Andrews, *Magic Squares and Cubes* (New York: Dover, 1960). A concise treatment can be found in A. S. Posamentier, B. S. Smith, and J. Stepelman, *Teaching Secondary School Mathematics: Techniques and Enrichment Units*, 8th ed. (Boston: Pearson/Allyn and Bacon, 2010), pp. 231–35.

**Figure 3-40**

| | | | |
|---|---|---|---|
| 16 | 3 | 2 | 13 |
| 5 | 10 | 11 | 8 |
| 9 | 6 | 7 | 12 |
| 4 | 15 | 14 | 1 |

**Figure 3-41**

We can see that the sum of all the rows and all the columns is 34. So that is all that would be required for this square matrix of numbers to be considered a "magic square." However, this "Dürer Magic Square" has many more properties that are not common to other magic squares. See if you can spot some of them.[16]

Elisha S. Loomis did not bypass the connection between the Pythagorean Theorem and magic squares.[17] He constructed some of his own—ones that he could connect with the Pythagorean Theorem. Consider the three magic squares shown in figure 3-42. The sum of each row or column of square I is 147. The sum of each row or column of square II is 46, and the sum of each row or column of square III is 125.

I

| | | |
|---|---|---|
| 48 | 53 | 46 |
| 47 | 49 | 51 |
| 52 | 45 | 50 |

II

| | | | |
|---|---|---|---|
| 4 | 18 | 17 | 7 |
| 15 | 9 | 10 | 12 |
| 11 | 13 | 14 | 8 |
| 16 | 6 | 5 | 19 |

III

| | | | | |
|---|---|---|---|---|
| 15 | 16 | 33 | 30 | 31 |
| 37 | 22 | 27 | 26 | 13 |
| 36 | 29 | 25 | 21 | 14 |
| 18 | 24 | 23 | 28 | 32 |
| 19 | 34 | 17 | 20 | 35 |

**Figure 3-42**

Notice that the sum of all the cells for each of the squares is

Square I: cell sum $= 3 \times 147 = 441$
Square II: cell sum $= 4 \times 46 = 184$
Square III: cell sum $= 5 \times 125 = 625$

16. A source that will provide you with some of these additional properties is Alfred S. Posamentier, *Math Charmers: Tantalizing Tidbits for the Mind* (Amherst, NY: Prometheus Books, 2002).

17. Elisha S. Loomis, *The Pythagorean Proposition* (Reston, VA: National Council of Teachers of Mathematics, 1968), p. 254.

The sum of the cells of square I + square II = 441 + 184 = 625, which is the sum of the cells of Square III.

Furthermore, assuming that all the cells are the same size, the sum of the areas of square I and square II equals the area of square III (9 + 16 = 25).

※※※

Applications of the Pythagorean Theorem in the realm of geometry are practically boundless. We have provided some highlights both in two and three dimensions (along with some recreational ideas). You may wish to extend some of the applications provided, or you may search for others. Your imagination is the only limitation here.

# Chapter 4

# Pythagorean Triples and Their Properties

We have found there are integers that can represent the sides of a right triangle. Some of the more common sets of three integers that can represent the lengths of the sides of a right triangle are (3, 4, 5), (5, 12, 13), and (7, 24, 25). That is, these triples satisfy the Pythagorean relationship $(a^2 + b^2 = c^2)$. We call an ordered set of three numbers $(a, b, c)$ that satisfies the Pythagorean Theorem $(a^2 + b^2 = c^2)$ a *Pythagorean triple*. The questions that often arise and that we hope to address in this chapter are: How many such triples are there? Is there a general way in which one can find these triples without just trying various combinations of three numbers to see if they satisfy the relationship? What are some properties of Pythagorean triples?

## Multiples of Pythagorean Triples

Our simplest Pythagorean triple is (3, 4, 5). Suppose we consider a multiple of this triple, such as (6, 8, 10), or, say, (15, 20, 25). Are these also Pythagorean triples?

Since $6^2 + 8^2 = 36 + 64 = 100 = 10^2$ and $15^2 + 20^2 = 225 + 400 = 625 = 25^2$, we can see that it would appear that multiples of a Pythagorean triple result in other Pythagorean triples. We can easily justify this conjecture algebraically as follows.

Suppose we consider the Pythagorean triple (3, 4, 5) and let ($3n$, $4n$, $5n$), where $n$ is a positive integer, be any multiple of this first triple. We now need to verify that ($3n$, $4n$, $5n$) is also a Pythagorean triple. We can do this as follows:

$$(3n)^2 + (4n)^2 = 9n^2 + 16n^2 = (9+16)n^2 = 25n^2 = (5n)^2$$

This verifies that ($3n$, $4n$, $5n$) is also a Pythagorean triple. This will allow us to conclude that there are an infinite number of Pythagorean triples that are multiples of (3, 4, 5).

Having established that there are infinitely many Pythagorean triples is not the whole picture, however, since all of the Pythagorean triples we have generated so far are all multiples of (3, 4, 5). Yet we know there are other Pythagorean triples that are not multiples of this triple, such as (5, 12, 13), (8, 15, 17), and (7, 24, 25), to name just a few. Each of these triples has no common factor (except 1), or we can say their members are relatively prime and they are called *primitive* Pythagorean triples. One is tempted to ask how many such primitive Pythagorean triples exist. As you might expect, there are an infinite number of such primitive Pythagorean triples. Let's investigate this further by considering various ways to generate Pythagorean triples.

## Fibonacci's Method for Finding Pythagorean Triples

Perhaps one of the most influential mathematicians of the thirteenth century was Leonardo of Pisa—better known today as Fibonacci. His fame today results from a mathematical problem in a book he first published in 1202 titled *Liber Abaci* (Book of Calculating). He presented a problem in chapter 12 of the book that involved the regeneration of rabbits. Its solution eventually led us to the now-famous Fibonacci numbers.[1] Also in this book he first introduced the Western world to the numerals we use today. In 1225, he published a book titled *Liber quadratorum* (Book of Squares) in which he stated the following:

> I thought about the origin of all square numbers and discovered that they arise out of the increasing sequence of odd numbers; for the unity is a square and from it is made the first square, namely, 1; that to this unity is added 3, making a second square, namely, 4, with root 2; if to the sum is added the third odd number, namely, 5, the third square is created, namely, 9, with root 3; and thus sums of consecutive odd numbers and a sequence of squares always arise together in order.

Fibonacci is essentially describing the following relationship:

$$\begin{aligned} 1 &= 1 = 1^2 \\ 1+3 &= 4 = 2^2 \\ 1+3+5 &= 9 = 3^2 \\ 1+3+5+7 &= 16 = 4^2 \\ 1+3+5+7+9 &= 25 = 5^2 \end{aligned}$$

In general terms the sum of the first $n$ odd numbers is $1+3+5+7+\ldots+(2n-1)=n^2$.

Notice how the squares in figure 4-1—beginning with the single square at the lower left—increase in area by the consecutive odd numbers just as we established algebraically. This is a geometric analog of this arithmetic statement.

---

1. For a more complete discussion of the Fibonacci numbers (1, 1, 2, 3, 5, 8, . . . , see Alfred S. Posamentier and Ingmar Lehmann, *The Fabulous Fibonacci Numbers* (Amherst, NY: Prometheus Books, 2007).

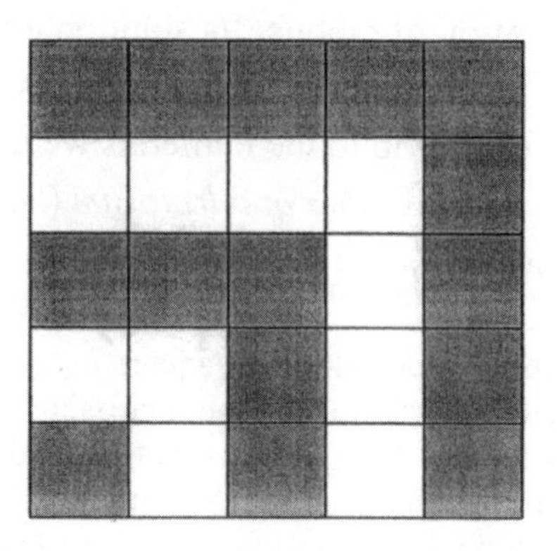

**Figure 4-1**

Although Fibonacci knew of the Pythagorean Theorem and therefore was aware of Pythagorean triples—after all, he lived about 1,700 years after Pythagoras—he was able to generate these triples in the following way. Let's consider the following series $1+3+5+7+9=5^2$, whose last term (9) is a perfect square. The sum we have enclosed here in parentheses $(1+3+5+7)$ is 16. So this equation can be rewritten as 16 + 9 = 25, which gives us the primitive Pythagorean triple (3, 4, 5).

Let's consider another series of consecutive odd integers—one that ends with a perfect square—to convince ourselves that this pattern can really generate other primitive Pythagorean triples.

$$1+3+5+7+9+11+13+15+17+19+21+23+25=169=13^2$$

Using the same procedure as above, we add the terms to the next to last one—here shown within parentheses:

$$(1+3+5+7+9+11+13+15+17+19+21+23)+25=169=13^2$$

which then can be written as $144+25=169$, or $12^2+5^2=13^2$. This gives us another primitive Pythagorean triple (5, 12, 13). We can continue this for an infinite number of such series that end in a perfect square, because there are an infinite number of odd perfect squares—since every odd number has an odd perfect square. We can then conclude that there are an infinite number of Pythagorean triples. However, even though this will generate an infinite number of Pythagorean triples, it does not generate all possible Pythagorean triples. You will notice a pattern[2] forming when you apply this to a number of such series of odd numbers. (See figure 4-2.)

| *a* | *b* | *c* |
|---|---|---|
| 3 | 4 | 5 |
| 5 | 12 | 13 |
| 7 | 24 | 25 |
| 9 | 40 | 41 |
| 11 | 60 | 61 |
| 13 | 84 | 85 |
| 15 | 112 | 113 |
| 17 | 144 | 145 |
| 19 | 180 | 181 |

**Figure 4-2**

## Euclid's Method for Finding Pythagorean Triples

The question then arises: How can we more succinctly generate primitive Pythagorean triples? More important, how can we obtain all Pythagorean triples? That is, is there a formula for achieving this goal? One such formula, attributed to the work of Euclid, generates values of $a$, $b$, and $c$, where $a^2 + b^2 = c^2$ as follows:

$$a = m^2 - n^2$$

$$b = 2mn$$

$$c = m^2 + n^2$$

We can easily show that these will always yield a Pythagorean triple by squaring each term of the potential triple:

2. For each case, $c - b = 1$.

$$a^2 = \left(m^2 - n^2\right)^2$$
$$b^2 = \left(2mn\right)^2$$
$$c^2 = \left(m^2 + n^2\right)^2$$

We will do this simple algebraic task by showing that the sum of $a^2 + b^2$ is actually equal to $c^2$.

$$a^2 + b^2 = \left(m^2 - n^2\right)^2 + \left(2mn\right)^2$$
$$a^2 + b^2 = m^4 - 2m^2n^2 + n^4 + 4m^2n^2$$
$$a^2 + b^2 = m^4 + 2m^2n^2 + n^4 = \left(m^2 + n^2\right)^2 = c^2$$

Therefore, $a^2 + b^2 = c^2$

## Applying Euclid's Formula to Gain an Insight into Pythagorean Triple Properties

If we apply this formula to some values of $m$ and $n$ in figure 4-3, we should notice a pattern as to when the triple will be primitive and also discover some other possible patterns.

An inspection of the triples in figure 4-3 might lead us to make the following conjecture—which, of course, can be proved. For example, the formula $a = m^2 - n^2$, $b = 2mn$, and $c = m^2 + n^2$ will yield *primitive* Pythagorean triples only when $m$ and $n$ are relatively prime—that is, when they have no common factor other than 1—and *exactly one* of these must be an even number, with $m > n$.

This formula will allow us to discover many relationships that exist among these Pythagorean triples. For example, when we inspect the values of $m$ and $n$ that determine a primitive Pythagorean triple and in addition where $n = 1$, we can notice that, as in figure 4-4, the hypotenuse will differ from one of the legs by 2.

| $m$ | $n$ | $a = m^2 - n^2$ | $b = 2mn$ | $c = m^2 + n^2$ | **Pythagorean triple $(a, b, c)$** | **Primitive** |
|---|---|---|---|---|---|---|
| 2 | 1 | 3 | 4 | 5 | (3, 4, 5) | Yes |
| 3 | 1 | 8 | 6 | 10 | (6, 8, 10) | No |
| 3 | 2 | 5 | 12 | 13 | (5, 12, 13) | Yes |
| 4 | 1 | 15 | 8 | 17 | (8, 15, 17) | Yes |
| 4 | 2 | 12 | 16 | 20 | (12, 16, 20) | No |
| 4 | 3 | 7 | 24 | 25 | (7, 24, 25) | Yes |
| 5 | 1 | 24 | 10 | 26 | (10, 24, 26) | No |
| 5 | 2 | 21 | 20 | 29 | (20, 21, 29) | Yes |
| 5 | 3 | 16 | 30 | 34 | (16, 30, 34) | No |
| 5 | 4 | 9 | 40 | 41 | (9, 40, 41) | Yes |
| 6 | 1 | 35 | 12 | 37 | (12, 35, 37) | Yes |
| 6 | 2 | 32 | 24 | 40 | (24, 32, 40) | No |
| 6 | 3 | 27 | 36 | 45 | (27, 36, 45) | No |
| 6 | 4 | 20 | 48 | 52 | (20, 48, 52) | No |
| 6 | 5 | 11 | 60 | 61 | (11, 60, 61) | Yes |
| 7 | 1 | 48 | 14 | 50 | (14, 48, 50) | No |
| 7 | 2 | 45 | 28 | 53 | (28, 45, 53) | Yes |
| 7 | 3 | 40 | 42 | 58 | (40, 42, 58) | No |
| 7 | 4 | 33 | 56 | 65 | (33, 56, 65) | Yes |
| 7 | 5 | 24 | 70 | 74 | (24, 70, 74) | No |
| 7 | 6 | 13 | 84 | 85 | (13, 84, 85) | Yes |
| 8 | 1 | 63 | 16 | 65 | (16, 63, 65) | Yes |
| 8 | 2 | 60 | 32 | 68 | (32, 60, 68) | No |
| 8 | 3 | 55 | 48 | 73 | (48, 55, 73) | Yes |
| 8 | 4 | 48 | 64 | 80 | (48, 64, 80) | No |
| 8 | 5 | 39 | 80 | 89 | (39, 80, 89) | Yes |
| 8 | 6 | 28 | 96 | 100 | (28, 96, 100) | No |
| 8 | 7 | 15 | 112 | 113 | (15, 112, 113) | Yes |

**Figure 4-3**

Algebraically this is easily demonstrated. Consider, again, the formula for generating all Pythagorean triples:

$$a = m^2 - n^2,\ b = 2mn, \text{ and } c = m^2 + n^2$$

When $n = 1$, we get $a = m^2 - 1^2 = m^2 - 1$ and $c = m^2 + 1^2 = m^2 + 1$. Therefore, the difference we find between $a$ and $c$ is $c - a = \left(m^2 + 1\right) - \left(m^2 - 1\right) = 2$. Figure 4-4 shows a few cases of primitive Pythagorean triples where $n = 1$.

| $m$ | $n$ | $a = m^2 - n^2$ | $b = 2mn$ | $c = m^2 + n^2$ | **Pythagorean triple ($a, b, c$)** | **Primitive** |
|---|---|---|---|---|---|---|
| 2 | 1 | 3 | 4 | 5 | (3, 4, 5) | Yes |
| 4 | 1 | 15 | 8 | 17 | (8, 15, 17) | Yes |
| 6 | 1 | 35 | 12 | 37 | (12, 35, 37) | Yes |
| 8 | 1 | 63 | 16 | 65 | (16, 63, 65) | Yes |
| 14 | 1 | 195 | 28 | 197 | (28, 195, 197) | Yes |
| 18 | 1 | 323 | 36 | 325 | (36, 323, 325) | Yes |
| 22 | 1 | 483 | 44 | 485 | (44, 483, 485) | Yes |

**Figure 4-4**

For which values of $m$ and $n$ will the hypotenuse differ from the larger leg by 1? Inspection of the various values listed in figure 4-3 reveals that when $m - n = 1$, then $c - b = 1$. This can be easily shown algebraically as follows.

We need to inspect $\left(m^2 + n^2\right) - \left(2mn\right)$ and see what happens when $m - n = 1$. First we find that $\left(m^2 + n^2\right) - \left(2mn\right) = \left(m - n\right)^2$. When $m - n = 1$, $\left(m - n\right)^2 = 1^2 = 1$, which verifies our conjecture about the difference of $c - b = 1$. This is then the second case of where the Pythagorean triples whose hypotenuse differs from a leg by 1 has gotten special attention.

## Pythagoras's Method for Finding Pythagorean Triples

The formula for finding Pythagorean triples that is attributed to Pythagoras is simple to use but doesn't generate all the triples, even though it will clearly generate an infinite number of triples. The formula specifies that if side $a$ is odd, then side $b = \frac{a^2 - 1}{2}$ and side $c = \frac{a^2 + 1}{2}$. We can apply this, for example, for (odd) side $a = 5$. Then $b = \frac{5^2 - 1}{2} = 12$ and $c = \frac{a^2 + 1}{2} = 13$, which gives us the

Pythagorean triple (5, 12, 13). You may wish to apply Pythagoras's formula to other odd values of side $a$. You should notice a pattern when you generate a few more triples with Pythagoras's formula, and you will also note some Pythagorean triples will not be included.

## Plato's Method for Finding Pythagorean Triples

Plato approached the problem of generating Pythagorean triples by beginning with an even number for side $a$. His formula for finding the remaining two members of the Pythagorean triple is: side $b=\left(\frac{a}{2}\right)^2-1$ and side $c=\left(\frac{a}{2}\right)^2+1$. Applying this formula to even side $a=6$, we get $b=\left(\frac{6}{2}\right)^2-1=8$ and side $c=\left(\frac{6}{2}\right)^2+1=10$, which gives us the Pythagorean triple (6, 8, 10)—although not a primitive Pythagorean triple!

Yet for the even side $a$ = 8, applying Plato's formula gives us $b=\left(\frac{8}{2}\right)^2-1=15$, and $c=\left(\frac{8}{2}\right)^2+1=17$, resulting in the Pythagorean triple (8, 15, 17), which is a primitive Pythagorean triple. You might want to determine which values of $a$ will yield primitive Pythagorean triples and which will result in nonprimitive Pythagorean triples.

## A Novel Method for Generating All Primitive Pythagorean Triples

We begin with any primitive Pythagorean triple, say, ($a$, $b$, $c$), and we will substitute these values of $a$, $b$, and $c$ into the three sets of formulas in figure 4-5. Each will generate a new primitive Pythagorean triple ($x$, $y$, $z$).

| | $x$ | $y$ | $z$ |
|---|---|---|---|
| **Formula 1** | $a-2b+2c$ | $2a-b+2c$ | $2a-2b+3c$ |
| **Formula 2** | $a+2b+2c$ | $2a+b+2c$ | $2a+2b+3c$ |
| **Formula 3** | $-a+2b+2c$ | $-2a+b+2c$ | $-2a+2b+3c$ |

**Figure 4-5**

To see how this works, we will apply the three formulas to the primitive Pythagorean triple (5, 12, 13).

| | $x$ | $y$ | $z$ | $(x, y, z)$ |
|---|---|---|---|---|
| **Formula 1** | $(5) - 2(12) + 2(13) = 7$ | $2(5) - (12) + 2(13) = 24$ | $2(5) - 2(12) + 3(13) = 25$ | (7, 24, 25) |
| **Formula 2** | $(5) + 2(12) + 2(13) = 55$ | $2(5) + (12) + 2(13) = 48$ | $2(5) + 2(12) + 3(13) = 73$ | (55, 48, 73) |
| **Formula 3** | $-(5) + 2(12) + 2(13) = 45$ | $-2(5) + (12) + 2(13) = 28$ | $-2(5) + 2(12) + 3(13) = 53$ | (45, 28, 53) |

**Figure 4-6**

Essentially, we can use any primitive Pythagorean triple beginning with the smallest (3, 4, 5) to generate others with these three formulas.

## A Fascinating Approach to Generate Pythagorean Triples

This may seem a bit contrived, but for rather simple reasons it does work. We begin by creating a sequence of mixed numbers of the following form:

$$1\frac{1}{3},\ 2\frac{2}{5},\ 3\frac{3}{7},\ 4\frac{4}{9},\ 5\frac{5}{11},\ 6\frac{6}{13},\ 7\frac{7}{15},\ \ldots,\ n\frac{n}{2n+1}$$

Here the whole number parts of the above mixed numbers are simply the natural numbers in order, the numerators of the fractions are the same number as the whole number, and the denominators of the fractions are consecutive odd numbers, beginning with 3. We now convert each of the fractions in this sequence to an improper fraction[3] to get the following sequence:

$$\frac{4}{3}, \frac{12}{5}, \frac{24}{7}, \frac{40}{9}, \frac{60}{11}, \frac{84}{13}, \frac{112}{15}, \ldots, \frac{n(2n+1)+n}{2n+1} = \frac{2n(n+1)}{2n+1}$$

The fractions (when reduced to lowest terms) will produce the first two members of a Pythagorean triple, allowing us to get the third member of the triple by simply applying the Pythagorean Theorem. For example, if we take the sixth term of this sequence, $\frac{84}{13}$, we have the first two members of the Pythagorean triple (13, 84, $c$). Then to get the third member, we simply find $c = \sqrt{84^2 + 13^2} = \sqrt{7,056 + 169} = 85$. Thus, the complete Pythagorean triple is (13, 84, 85). To help you understand this method we will apply it once more. This time we will select the fraction $\frac{60}{11}$ from this sequence. Again, the Pythagorean triple is (11, 60, $c$). To find the value of $c$, we apply the Pythagorean Theorem to get $\sqrt{11^2 + 60^2} = \sqrt{121 + 3,600} = 61$. You might want to try a few other examples to convince yourself that this does generate Pythagorean triples. Notice that these Pythagorean triples all seem to have the larger leg differ from the hypotenuse by 1.

In order to justify that this will always be true, we can use some elementary algebra, beginning with the general term above $\frac{2n(n+1)}{2n+1}$ and show that the sum of the numerator squared and the denominator squared results in a perfect square as well. That is, $(2n(n+1))^2 + (2n+1)^2$ ought to be a perfect square.

3. An improper fraction is one where the numerator is greater than the denominator.

Let us simplify this expression: $(2n(n+1))^2+(2n+1)^2$

$$(2n(n+1))^2+(2n+1)^2=(4n^4+8n^3+4n^2)+(4n^2+4n+1)$$
$$4n^4+8n^3+8n^2+4n+1=(2n^2+2n+1)^2$$

And so we have shown that the sum of the squares of the numerator and denominator of these fractions is a perfect square.

## Fibonacci's (Indirect) Connection with the Pythagorean Triples

As mentioned earlier, the sequence that has made Fibonacci so famous today was in chapter 12 of his book *Liber Abaci* (1202) and has to do with the regeneration of rabbits. The famous sequence named after him is 1, 1, 2, 3, 5, 8, 13, 21, 34, 55, 89, 144, … , where after the first two 1s each term is the sum of its two predecessors. On the surface, it would appear that there is nothing here in common with the Pythagorean triples. Well, we have a surprise for you: one can generate Pythagorean triples from this completely unrelated sequence of numbers. Interestingly, they were discovered independently and have no subsequent connection.

To create a Pythagorean triple from the Fibonacci numbers, we take any four consecutive numbers in this sequence, such as 3, 5, 8, and 13. These four Fibonacci numbers are the fourth through the seventh Fibonacci numbers, often designated by $F_4$, $F_5$, $F_6$, and $F_7$. We now follow these rules:[4]

4. A proof of this delightful (and surprising) procedure can be found in appendix A.

1. Multiply the middle two numbers and double the result

   Here the product of 5 and 8 is 40; then we double this to get **80** (This is one member of the Pythagorean triple.)

2. Multiply the two outer numbers

   Here the product of 3 and 13 is **39**
   (This is another member of the Pythagorean triple.)

3. Add the squares of the inner two numbers to get the third member of the Pythagorean triple

   Here $5^2 + 8^2 = 25 + 64 = \mathbf{89}$

So we have found a Pythagorean triple: (39, 80, 89). We can verify that this is, in fact, a Pythagorean triple by showing that $39^2 + 80^2 = 1{,}521 + 6{,}400 = 7{,}921 = 89^2$.

While we are admiring this relationship, we can take it a step further. The eleventh Fibonacci number ($F_{11}$) is 89. Now referring back to our original Fibonacci numbers, $F_4$, $F_5$, $F_6$, and $F_7$, we find that the sum of the indices[5] (5 and 6) of the two inner Fibonacci numbers and the sum of the indices of the two outer (4 and 7) Fibonacci numbers is the index of the hypotenuse we just found (11). Furthermore, the area of the Pythagorean triangle we found (39, 80, 89), namely, $\frac{1}{2} \cdot 39 \cdot 80 = 1{,}560$, also happens to equal $3 \cdot 5 \cdot 8 \cdot 13$, which is the product of our originally used Fibonacci numbers. Quite surprisingly unexpected!

You might want to convince yourself that this truly works for any four consecutive Fibonacci numbers by applying these rules elsewhere in the sequence.

---

5. The *index* in this case refers to the subscript of *F*.

## Another Connection between the Fibonacci Numbers and the Pythagorean Triples

Recall Euclid's formula for generating primitive Pythagorean triples: $a = m^2 - n^2$, $b = 2mn$, and $c = m^2 + n^2$, where $m > n$ and $m$ and $n$ are relatively prime natural numbers, with exactly one of them being even.

Recall, also, the Fibonacci numbers: 1, 1, 2, 3, 5, 8, 13, 21, 34, 55, 89, 144, … .

Suppose we let $m$ and $n$ take on the values of the Fibonacci numbers sequentially, beginning with $m = 2$ and $n = 1$. You will find, to your amazement, that in each case $c$ will also turn out to be a Fibonacci number, as you can see in figure 4-7.

| $k$ | $m$ | $n$ | $a = m^2 - n^2$ | $b = 2mn$ | $c = m^2 + n^2$ |
|---|---|---|---|---|---|
| 1 | 2 | 1 | 3 | 4 | 5 |
| 2 | 3 | 2 | 5 | 12 | 13 |
| 3 | 5 | 3 | 16 | 30 | 34 |
| 4 | 8 | 5 | 39 | 80 | 89 |
| 5 | 13 | 8 | 105 | 208 | 233 |
| 6 | 21 | 13 | 272 | 546 | 610 |
| 7 | 34 | 21 | 715 | 1428 | 1597 |
| 8 | 55 | 34 | 1869 | 3740 | 4181 |
| 9 | 89 | 55 | 4896 | 9790 | 10946 |
| 10 | 144 | 89 | 12815 | 25632 | 28657 |
| 11 | 233 | 144 | 33553 | 67104 | 75025 |
| 12 | 377 | 233 | 87840 | 175682 | 196418 |
| … | … | … | … | … | … |

**Figure 4-7**

We can easily justify this phenomenon. Consider the general case for two consecutive Fibonacci numbers, where $F_n$ is the $n$th Fibonacci number. We therefore let $m_k = F_{k+2}$ and $n_k = F_{k+1}$.

Using Euclid's formulas (above), we get

$$a_k = m^2 - n^2 = (F_{k+2})^2 - (F_{k+1})^2$$

$$b_k = 2mn = 2F_{k+2}\,F_{k+1}$$

$$c_k = m^2 + n^2 = (F_{k+2})^2 + (F_{k+1})^2$$

We can see that the sum of the squares of two consecutive Fibonacci numbers will always be another Fibonacci number $F_{2k+1}$.[6]

## Triangular Numbers and the Pythagorean Triples

First let us recall what the triangular numbers are. They are numbers such as 1, 3, 6, 10, 15, 21, 28, ... that represent the number of dots that can be placed in the arrangement of an equilateral triangle, as shown in figure 4-8.

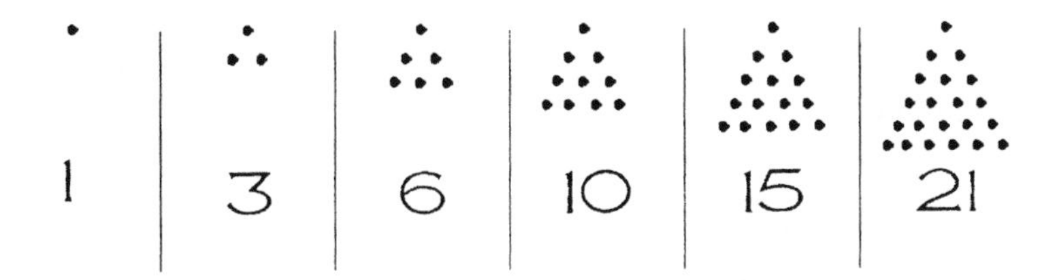


**Figure 4-8**

You will notice that the differences between the consecutive triangular numbers form a sequence, 2, 3, 4, 5, 6, 7, . . . . Or put another way, any triangular number, $t_n$, can be expressed as the sum of the first $n$ consecutive natural numbers. For example, 1 + 2 + 3 + 4 + 5 + 6 + 7 = 28, which is the seventh triangular number, $t_7$. In the general case, we can express the $n$th triangular number as $t_n = \frac{n(n+1)}{2}$. Furthermore, the sum of any two consecutive triangular numbers is always a square number, as, for example, 15 + 21 = 36. This can be shown geometrically as in figure 4-9.

6. A proof of this conclusion can be found in appendix A.

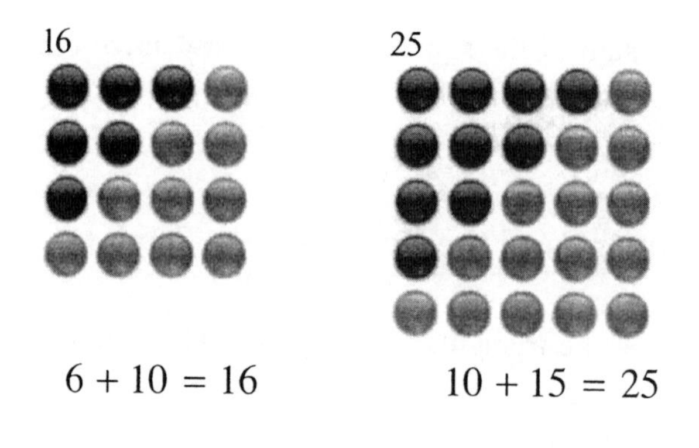


**Figure 4-9**

Some triangular numbers have an additional property; namely, they can be expressed as the product of three consecutive numbers. For example,

$$t_3 = 6 = 1 \cdot 2 \cdot 3,$$
$$t_{15} = 120 = 4 \cdot 5 \cdot 6$$
$$t_{20} = 210 = 5 \cdot 6 \cdot 7$$
$$t_{44} = 990 = 9 \cdot 10 \cdot 11$$

Some other triangular numbers that can be expressed as the product of three consecutive numbers are $t_{608} = 185{,}136$ and $t_{22{,}736} = 258{,}474{,}216$. By the way, a nice pattern with triangular numbers is that numbers of the form 21, 2211, 222111, ... are all triangular![7]

By now you might be wondering what the triangular numbers have to do with the Pythagorean triples. In figure 4-10 we list the Pythagorean triples, where $a$ is odd (consecutively). An inspection of the $b$ val-

7. This can be verified by showing that the general term for a triangular number, $\frac{n(n+1)}{2}$, holds true for each of these numbers. For example,

$$21 = \frac{6 \cdot 7}{2},\ 2{,}211 = \frac{66 \cdot 67}{2},\ \text{and}\ 222{,}111 = \frac{666 \cdot 667}{2}$$

ues will reveal a surprising pattern. To discover this pattern we need to factor the *b*-value entries. (See figure 4-11.)

| *a* | *b* | *c* |
|---|---|---|
| 3 | 4 | 5 |
| 5 | 12 | 13 |
| 7 | 24 | 25 |
| 9 | 40 | 41 |
| 11 | 60 | 61 |
| 13 | 84 | 85 |
| 15 | 112 | 113 |
| 17 | 144 | 145 |
| 19 | 180 | 181 |
| 21 | 220 | 221 |
| 23 | 264 | 265 |
| 25 | 312 | 313 |

**Figure 4-10**

The triangular numbers are embedded in the *b* values. If we, now, just analyze the *b* values in figure 4-11, you should notice the consecutive triangular numbers appearing.

Believe it or not, there is even a Pythagorean triple made up entirely of triangular numbers: (8,778; 10,296; 13,530), where 8,778 is the 132nd triangular number, 10296 is the 143rd triangular number, and 13,530 is the 164th triangular number.

| *b* |
|---|
| $4 = 4 \times \mathbf{1}$ |
| $12 = 4 \times \mathbf{3}$ |
| $24 = 4 \times \mathbf{6}$ |
| $40 = 4 \times \mathbf{10}$ |
| $60 = 4 \times \mathbf{15}$ |
| $84 = 4 \times \mathbf{21}$ |
| $112 = 4 \times \mathbf{28}$ |
| $144 = 4 \times \mathbf{36}$ |
| $180 = 4 \times \mathbf{45}$ |
| $220 = 4 \times \mathbf{55}$ |
| $264 = 4 \times \mathbf{66}$ |
| $312 = 4 \times \mathbf{78}$ |

**Figure 4-11**

## Consecutive Members of a Pythagorean Triple

When we further inspect the list of Pythagorean triples in figure 4-10 we notice, not only that $c = b + 1$, but also that $a^2 = b + c$, a truly remarkable pattern among these selected Pythagorean triples, which were originally selected because their $a$ values are consecutive odd numbers. For example, for the Pythagorean triple (7, 24, 25) we have $25 = 24 + 1$, and at the same time we also have $7^2 = 24 + 25 = 49$.

There are further gems embedded in these special Pythagorean triples. Let's look at the Euclidean formula values of $m$ and $n$ for just these Pythagorean triples in figure 4-12.

Notice the values of $m$ and $n$ in this table. Each of them is a pair of consecutive natural numbers—maintaining the requirement for primitive Pythagorean triples that $m > n$—and since they are consecutive, we know they cannot have a common factor so they fit the other requirement for primitive Pythagorean triples that they be relatively prime, that is, that they have no common factor other than 1 and one will always be even.

| $m$ | $n$ | $a = m^2 - n^2$ | $b = 2mn$ | $c = m^2 + n^2$ |
|---|---|---|---|---|
| 2 | 1 | 3 | 4 | 5 |
| 3 | 2 | 5 | 12 | 13 |
| 4 | 3 | 7 | 24 | 25 |
| 5 | 4 | 9 | 40 | 41 |
| 6 | 5 | 11 | 60 | 61 |
| 7 | 6 | 13 | 84 | 85 |
| 8 | 7 | 15 | 112 | 113 |
| 9 | 8 | 17 | 144 | 145 |
| 10 | 9 | 19 | 180 | 181 |
| 11 | 10 | 21 | 220 | 221 |
| 12 | 11 | 23 | 264 | 265 |
| 13 | 12 | 25 | 312 | 313 |

**Figure 4-12**

In case you are wondering if it is always true that if $m$ and $n$ are consecutive numbers that the hypotenuse size will differ from a leg by 1, we will provide a very simple algebraic justification (or proof).

When $m$ and $n$ are consecutive numbers, we can say $m = n + 1$. Substituting for $m$ in Euclid's formula:

$$a = m^2 - n^2 = (n+1)^2 - n^2 = 2n+1$$
$$b = 2mn = 2(n+1)n = 2n^2 + 2n$$
$$c = m^2 + n^2 = (n+1)^2 + n^2 = 2n^2 + 2n + 1$$

You can then see that $c = b+1$. The ambitious reader may question if there are other primitive Pythagorean triples where the hypotenuse is 1 greater than a leg, ones which are not included in the pattern we established in figure 4-12—that is, where $c = a+1$—and that do not have the relationship of $m = n+1$.

It can be easily shown that there are no other Pythagorean triples beyond those that follow this pattern. We can demonstrate this rather simply. Assume that there are Pythagorean triples where $c = a+1$, which we can write as $m^2 + n^2 = \left(m^2 - n^2\right) + 1$. This results in $n^2 = -n^2 + 1$, or $n^2 = \frac{1}{2}$, which is not possible since $n$ is an integer. Therefore our assumption, that $c = a+1$, is wrong; so $c \neq a+1$, or the hypotenuse can never be 1 greater than the other leg.

Now let's see what happens if we assume that $c = b+1$, or that $m^2 + n^2 = 2mn + 1$. We can write this as $m^2 - 2mn + n^2 = 1$, or $\left(m-n\right)^2 = 1$. This leads us to $m - n = 1$, or $m - n = -1$. However, $m - n$ cannot be negative, since $m > n$. Therefore, we have $m - n = 1$ or $m = n+1$, which tells us that there are no other Pythagorean triples other than those in the pattern shown in figure 4-12, where the hypotenuse is 1 greater than a leg.

One more word about the $b$ terms (i.e., $2mn$) in this pattern-rich list in figure 4-12, namely, 4, 12, 24, 40, 60, 84, .... They just happen to fold nicely into the following pattern:

$$3^2 + \left[4^2\right] = 5^2$$
$$10^2 + 11^2 + \left[12^2\right] = 13^2 + 14^2$$
$$21^2 + 22^2 + 23^2 + \left[24^2\right] = 25^2 + 26^2 + 27^2$$
$$36^2 + 37^2 + 38^2 + 39^2 + \left[40^2\right] = 41^2 + 42^2 + 43^2 + 44^2$$
$$55^2 + 56^2 + 57^2 + 58^2 + 59^2 + \left[60^2\right] = 61^2 + 62^2 + 63^2 + 64^2 + 65^2$$

Are there many Pythagorean triples that are three consecutive numbers, such as (3, 4, 5)? To search for such triples we let the three members of the triple be $p-1, p,$ and $p+1$. Now testing them against the Pythagorean Theorem relationship we get

$$(p-1)^2 + p^2 = (p+1)^2$$
$$p^2 - 2p + 1 + p^2 = p^2 + 2p + 1$$
$$p^2 = 4p$$
$$p = 4$$

The resulting Pythagorean triple is then (3, 4, 5), where the three members are in consecutive order (i.e., each of the members differing by 1). Moreover, we should note that this is the only Pythagorean triple where the three members are consecutive. There is a simple way to show this to be true. Suppose we say that

$$m^2 - n^2 = X;$$
$$2mn = X + 1; \text{ and}$$
$$m^2 + n^2 = X + 2$$

thus making the three members of the Pythagorean triple consecutive numbers—through Euclid's formula. Then by adding:

$$\left(m^2 - n^2\right) + \left(m^2 + n^2\right) = X + X + 2;$$
$$2m^2 = 2X + 2; \text{ and}$$
$$m^2 = X + 1$$

and by subtracting:

$$\left(m^2 + n^2\right) - \left(m^2 - n^2\right) = X + 2 - X;$$
$$2n^2 = 2; \text{ and}$$
$$n^2 = 1$$

Since both $2mn$ and $m^2$ are equal to $X + 1$, they are equal; therefore, $2n = m$. However, since $n^2 = 1$, $n = 1$, then $m = 2$. These values of *m* and *n* give us only the (3, 4, 5) Pythagorean triple.

Let us turn our attention now to the possibility of having two consecutive numbers representing the two legs of a Pythagorean triple. Our most familiar Pythagorean triple, (3, 4, 5), already meets this criterion, since 3 and 4 are consecutive numbers. Which other Pythagorean triples have consecutive legs? We can take, for example, the Pythagorean triple (20, 21, 29), which clearly meets this criterion. Are there others? We can inspect the list of Pythagorean triples provided in appendix C, where we will find the following triples, where the legs are consecutive integers. (See figure 4-13.)

| $m$ | $n$ | $a = m^2 - n^2$ | $b = 2mn$ | $c = m^2 + n^2$ |
|---|---|---|---|---|
| 2 | 1 | 3 | 4 | 5 |
| 5 | 2 | 21 | 20 | 29 |
| 12 | 5 | 119 | 120 | 169 |
| 29 | 12 | 697 | 696 | 985 |
| 70 | 29 | 4059 | 4060 | 5741 |
| 169 | 70 | 23661 | 23660 | 33461 |
| 408 | 169 | 137903 | 137904 | 195025 |
| 985 | 408 | 803761 | 803760 | 1113689 |
| 2378 | 985 | 4684659 | 4684660 | 6625109 |
| 5741 | 2378 | 27304197 | 27304196 | 38613965 |

**Figure 4-13**

Again, if you search for it, a pattern can be found—as is so often the case in mathematics. However, this pattern is a bit different from earlier ones. After starting with a Pythagorean triple with consecutive number legs, to get the next Pythagorean triple where the two legs differ by 1 (i.e., they are consecutive numbers), we can do the following:

Let us begin with the Pythagorean triple (119, 120, 169).

To get the smallest leg of the next triple:

- Multiply the smaller leg from the previous triple by 6:
    - Example: $6 \times 119 = 714$
- Then subtract the smaller leg from the triple two before the one we seek:
    - Example: $714 - 20 = 694$
- Then add 2:
    - Example: $694 + 2 = 696$

To get the other leg of the Pythagorean triple, we just have to add 1, since we were searching for triples with consecutive integer legs. The third side can then be found by applying the Pythagorean Theorem!

Before leaving this list of Pythagorean triples (figure 4-13), you might want to admire some of the other patterns that exist here; namely, the $m$ value of one triple becomes the $n$ value of the next triple. Also, every other $m$ (or $n$) value is the hypotenuse of an earlier Pythagorean triple.

## Which Numbers Can Be Members of a Pythagorean Triple?

We confidently consider the Pythagorean triple (3, 4, 5) as the smallest-value Pythagorean triple. Clearly, 1 and 2 cannot be the legs of a Pythagorean triple. Try using these numbers and you will see the dilemma.

Let's see if we can determine a systematic way to find out if all integers greater than 2 can be represented in some Pythagorean triple.

We begin by inspecting odd integers greater than 1. We will let $a = \frac{1}{2}n$, and delete the remainder. We can then use the procedure shown in the section on "A Fascinating Approach to Generate Pythagorean Triples" (pages 132–34) to find a Pythagorean triple with that odd-number leg. For example, if we choose $a = 23$, and take half, ignoring the remainder, we get 11. Then, using this procedure, we form the mixed number $11\frac{11}{23} = \frac{264}{23}$, which determines the Pythagorean triple (23, 264, 265). In the same fashion we can get every odd number greater than 1 to represent a leg of a Pythagorean triple.

We now have to inspect the possible even numbers that can be part of a Pythagorean triple. First we shall consider those even numbers that are not powers of 2. By continuously halving this number, you will eventually end up with an odd number. We just established that every odd number greater than 1 will be a member of some Pythagorean triple. Therefore, by multiplying back by twos we can show that this nonpower of 2 can represent a member of a Pythagorean triple. For example, let's try to see if 28 can represent a member of a Pythagorean triple. We divide by 2 successively to get (28, 14, 7). Our result, 7, is an odd number and, therefore, as we just established (above), it is a member of a Pythagorean triple—in this case (7, 24, 25). So if we now multiply by the twos we divided by before, that is, twice, we get (28, 96, 100), a Pythagorean triple, although not a primitive one.

All that remains for us to establish now in seeking to find what numbers can be members of a Pythagorean triple is seeing if a number that is a power of 2 can be a member of a Pythagorean triple. If we begin with the Pythagorean triple (3, 4, 5) and keep doubling its members, the second member will be able to assume all powers of 2. For example, if we want to determine if 32 could be a member of a Pythagorean triple, then dividing by 2 continuously we will eventually reach 4, the member of the Pythagorean triple (3, 4, 5). Therefore, multiplying this triple's members by 8 gives us the Pythagorean triple (24, 32, 40), and we have shown that 32 can be a member of a Pythagorean triple. Thus, we have shown that all natural numbers greater than 2 can be members of some Pythagorean triple—primitive or not.

Furthermore, it is interesting to note how many Pythagorean triples have the various powers of 2 as a member. The listing in figure 4-14 should present the answer. Notice that the number of Pythagorean triples is 1 less than the exponent of 2.

| Pythagorean Triple Member | Pythagorean Triples | Number of Pythagorean Triples |
|---|---|---|
| $2 = 2^1$ | none | 0 |
| $4 = 2^2$ | (3, 4, 5) | 1 |
| $8 = 2^3$ | (6, 8, 10), (8, 15, 17) | 2 |
| $16 = 2^4$ | (12, 16, 20), (16, 30, 34), (16, 63, 65) | 3 |
| $32 = 2^5$ | (24, 32, 40), (32, 60, 68), (32, 126, 130) (32, 255, 257) | 4 |
| ⋮ | ⋮ | ⋮ |
| $2^{n+1}$ | | $n$ |

**Figure 4-14**

| Triple | Perimeter | Area | Triple | Perimeter | Area | Triple | Perimeter | Area |
|---|---|---|---|---|---|---|---|---|
| 3, 4, 5 | 12 | 6 | 27, 36, 45 | 108 | 486 | 24, 70, 74 | 168 | 840 |
| 6, 8, 10 | 24 | 24 | 30, 40, 50 | 120 | 600 | 45, 60, 75 | 180 | 1,350 |
| 5, 12, 13 | 30 | 30 | 14, 48, 50 | 112 | 336 | 21, 72, 75 | 168 | 756 |
| 9, 12, 15 | 36 | 54 | 24, 45, 51 | 120 | 540 | 30, 72, 78 | 180 | 1,080 |
| 8, 15, 17 | 40 | 60 | 20, 48, 52 | 120 | 480 | 48, 64, 80 | 192 | 1,536 |
| 12, 16, 20 | 48 | 96 | 28, 45, 53 | 126 | 630 | 18, 80, 82 | 180 | 720 |
| 15, 20, 25 | 60 | 150 | 33, 44, 55 | 132 | 726 | 51, 68, 85 | 204 | 1,734 |
| 7, 24, 25 | 56 | 84 | 40, 42, 58 | 140 | 840 | 40, 75, 85 | 200 | 1,500 |
| 10, 24, 26 | 60 | 120 | 36, 48, 60 | 144 | 864 | 36, 77, 85 | 198 | 1,386 |
| 20, 21, 29 | 70 | 210 | 11, 60, 61 | 132 | 330 | 13, 84, 85 | 182 | 546 |
| 18, 24, 30 | 72 | 216 | 39, 52, 65 | 156 | 1,014 | 60, 63, 87 | 210 | 1,890 |
| 16, 30, 34 | 80 | 240 | 25, 60, 65 | 150 | 750 | 39, 80, 89 | 208 | 1,560 |
| 21, 28, 35 | 84 | 294 | 33, 56, 65 | 154 | 924 | 54, 72, 90 | 216 | 1,944 |
| 12, 35, 37 | 84 | 210 | 16, 63, 65 | 144 | 504 | 35, 84, 91 | 210 | 1,470 |
| 15, 36, 39 | 90 | 270 | 32, 60, 68 | 160 | 960 | 57, 76, 95 | 228 | 2,166 |
| 24, 32, 40 | 96 | 384 | 42, 56, 70 | 168 | 1,176 | 65, 72, 97 | 234 | 2,340 |
| 9, 40, 41 | 90 | 180 | 48. 55, 73 | 176 | 1,320 | 60, 80, 100 | 240 | 2,400 |
| | | | | | | 28, 96, 100 | 224 | 1,344 |

**Figure 4-15**

## Areas and Perimeters of Pythagorean Triangles

We will now consider the Pythagorean triples from the point of view of the triangles they represent. The area of each right triangle is $\frac{ab}{2}$, and the perimeter is $a+b+c$. Figure 4-15 lists all the Pythagorean triangles with side lengths up to 100. An inspection of this list reveals some interesting facts about these right triangles:

- Only one has an area that is numerically less than its perimeter:
  - (3, 4, 5): the perimeter is 12 and the area is 6.
- Only two have their area numerically equal to their perimeter:
  - (6, 8, 10): the area and the perimeter are 24.
  - (5, 12, 13): the area and the perimeter are 30.

- Three triangles have an area that is numerically twice its perimeter:
    - (12, 16, 20): the area is 96 and the perimeter is 48.
    - (10, 24, 26): the area is 120 and the perimeter is 60.
    - (9, 40, 41): the area is 360 and the perimeter is 90.

There are many intriguing facts you can find on this list of Pythagorean triples—some of which you might want to test for "uniqueness." For example, the smallest common perimeter for two triangles among the Pythagorean triangles is 60, and for three triangles it is 120. In appendix D, we have an extended list of the Pythagorean triples. There you will find that for four triangles the smallest common perimeter is 360, and for five triangles it is 420. You might now seek other relationships among the perimeters of the Pythagorean triangles. A further entertaining activity would be to search for patterns and relationships among the areas of these right triangles.

For example, there are two right triangles listed in figure 4-15 with integer sides (i.e., a Pythagorean triple) that have an area of 210, namely, (20, 21, 29) and (12, 35, 37). Then going to an extended list, as in appendix D, we can find three right triangles with integer sides that have the same area[8] of 840. They are (40, 42, 58), (24, 70, 74), and (15, 112, 113). The smallest area common to three primitive Pythagorean triples is 13,123,110, and it comes from triangles with sides of lengths (4,485; 5,852; 7,373), (19,019; 1,380; 19,069), and (3,059; 8,580; 9,109).

---

8. Here are three other sets of three Pythagorean triples that have the same area:

(9,984; 48,160; 49,184), (26,880; 17,888; 32,288), and (7,280, 66,048; 66,448) —each has an area of 240,414,720;

(86,016; 48,640; 98,816), (26,880; 155,648; 157,952), and (20,480; 204,288; 205,312)—each has an area of 2,091,909,120; and

(1,069,929; 4,427,280; 4,554,729), (2,490,840; 1,901,718; 3,133,818), and (748,440; 6,328,998; 6,373,098)—each has an area of 2,368,437,631,560.

You may then want to extend your search for patterns or relationships involving the two—areas and perimeters—to find even more fascinating relationships.

※※※

Here is a rather unusual relationship—just to indicate the breadth of patterns to be found: For every natural number, $n$, there is at least one primitive Pythagorean triangle where numerically its area is $n$ times the perimeter. You can see this in figure 4-16 for the first seven of these primitive Pythagorean triples and you may want to search for the next few in the listing in appendix D.

| $n$ | **Pythagorean Triple** | **Perimeter** | **Area** |
|---|---|---|---|
| 1 | 5, 12, 13 | 30 | 30 |
| 2 | 9, 40, 41 | 90 | 180 |
| 3 | 20, 21, 29 | 70 | 210 |
| 4 | 17, 144, 145 | 306 | 1224 |
| 5 | 28, 45, 53 | 126 | 630 |
| 6 | 33, 56, 65 | 154 | 924 |
| 7 | 36, 77, 85 | 198 | 1386 |

**Figure 4-16**

By inspecting the listing of Pythagorean triples and the areas of their right triangles, you will notice that there is no area (i.e., $\frac{ab}{2}$, or by Euclid's formula $mn\left(m^2 - n^2\right)$) that is a perfect square. The great French mathematician Pierre de Fermat (1601–1665) proved that there is no Pythagorean triangle whose area is a perfect square. Therefore, you do not have to search any further, since we now know that no such area exists. Yet we can at least say that every area is divisible by 6. This is another peculiarity of Pythagorean triples.

Ambitious readers might also want to prove some of these relationships, which from observations are merely "conjectures" until they are proved true for all cases, making them valid generalizations.

## Some Other Pythagorean Curiosities

### *Curiosity 1*

Inspection of the list of Pythagorean triples will convince you that one of the first two members is always a multiple of 3, and one (possibly the same one) is always a multiple of 4. Therefore, the product of the two smaller members is always a multiple of 12, which makes the area of the right triangle with these sides always a multiple of 6. One member of the Pythagorean triple is always a multiple of 5. Therefore, the product of any Pythagorean triple is a multiple of 60.

### *Curiosity 2*

Whether there are two Pythagorean triples (primitive or nonprimitive) with the same product of its members is still an unanswered question. Having unanswered questions in mathematics is not unusual. Sometimes these questions get answered after many years of mathematical struggles. An example of a still-unsolved problem is known as the "Goldbach Conjecture," which was posed by Christian Goldbach (1690–1764). It states that every even number greater than 2 can be expressed as the sum of two prime numbers. There have been countless examples found, such as $12 = 5 + 7$, or $48 = 19 + 29$, and no one has ever found a counterexample. Yet since to date no one has proved this statement true for all cases, it remains only a *conjecture* and not a theorem. Thus even in the realm of the Pythagorean Theorem there are unsolved problems such as the one mentioned above.

On the other hand, Pierre de Fermat posed a challenge in 1643 and then answered it himself. He sought a Pythagorean triple where the sum of the two legs is a square integer and the hypotenuse is also a square integer. Symbolically, he sought to find a Pythagorean triple ($a$, $b$, $c$), where $a+b=p^2$ and $c=q^2$, where $p$ and $q$ are integers. He calculated one such Pythagorean triple to be (4,565,486,027,761; 1,061,652,293,520; 4,687,298,610,289), where the values $a + b$ = 4,565,486,027,761 + 51,061,652,293,520 = 627,138,321,281 = $2{,}372{,}159^2$. The hypotenuse is also a perfect square: $c$ = 4,687,298,610,289 = $2{,}165{,}017^2$. Fermat proved that this Pythagorean triple was the *smallest* having this property! It is hard to imagine the next larger such Pythagorean triple.

*Curiosity 3*

We can generate a "family" of rather unusual Pythagorean triples by using the formulas that we encountered on page 141:

$$a = 2n+1$$
$$b = 2n(n+1)$$
$$c = 2n(n+1)+1$$

| $n$ | $a = 2n+1$ | $b = 2n(n+1)$ | $c = 2n(n+1)+1$ |
|---|---|---|---|
| 10 | 21 | 220 | 221 |
| $10^2$ | 201 | 20200 | 20201 |
| $10^3$ | 2001 | 2002000 | 2002001 |
| $10^4$ | 20001 | 200020000 | 200020001 |
| $10^5$ | 200001 | 20000200000 | 20000200001 |
| $10^6$ | 2000001 | 2000002000000 | 2000002000001 |

**Figure 4-17**

You will be pleasantly surprised when you generate a similar list with powers of 20, 40, and so on, in place of the powers of 10 we used in figure 4-17. We can start you off by telling you that for $n = 20$, you will get $41^2 + 840^2 = 841^2$, and for $20^2$ you will get $401^2 + 80{,}400^2 = 80{,}401^2$. See what other patterns of this kind you can discover.

*Curiosity 4*

We can show that there are an infinite number of primitive Pythagorean triples where the third member (i.e., the hypotenuse) is the square of a natural number. This demonstration is rather simple. We can take any of the infinite primitive Pythagorean triples, say, $(n, m, h)$. By the Euclidean formula, where $x > y$, and $x$ and $y$ are relatively prime and of different parity (i.e., one is odd and the other is even):

$$n = x^2 - y^2$$
$$m = 2xy$$
$$h = x^2 + y^2$$

Since $(n, m, h)$ is a primitive triple, $n$ and $m$ are relatively prime and $m$ ($= 2xy$) is clearly even, so we can obtain a second primitive Pythagorean triple, $(a, b, c)$, by using the Euclidean formula[9] a second time.

Here we notice that by substituting the $x$ and $y$ values of $m$ and $n$ shown above, we get the following:

[9] The Euclidean formula is

$$a = m^2 - n^2$$
$$b = 2mn$$
$$c = m^2 + n^2$$

$$m^2 + n^2 = (2xy)^2 + (x^2 - y^2)^2 = 4x^2y^2 + x^4 - 2x^2y^2 + y^4$$
$$= x^4 + 2x^2y^2 + y^4 = (x^2 + y^2)^2$$

Thus we have shown that the third member of a Pythagorean triple, $m^2 + n^2$, will be a perfect square whenever we use the first two members of another primitive Pythagorean triple in the Euclidean formula to generate this second Pythagorean triple. Then this second Pythagorean triple has its third member (the hypotenuse) as a perfect square of a natural number.

We can apply this to a few primitive Pythagorean triples so you can see how what we just proved can be applied. Let's begin with our smallest primitive Pythagorean triple, (3, 4, 5). If we get the sum of the squares of the first two members, $3^2 + 4^2 = 25$, we have the hypotenuse (or the third member) of another primitive Pythagorean triple, (7, 24, 25). As another illustration of this relationship we could use the primitive Pythagorean triple we just obtained: (7, 24, 25). Again, we take the sum of the squares of the first two members: $7^2 + 24^2 = 49 + 576 = 625$, which is a perfect square ($25^2$). This is the hypotenuse of the primitive Pythagorean triple (175, 600, 625). From the primitive Pythagorean triple (8, 15, 17) we can generate the primitive Pythagorean triple (161, 240, 289) that has as its third member, 289, a perfect square ($17^2$). You might want to see which squares appear in the hypotenuse position of a primitive Pythagorean triple and see if some pattern evolves.

*Curiosity 5*

In a similar fashion we can also show that there are infinitely many primitive Pythagorean triples where one of the first two members is a perfect square. An example where the odd member is a square is (9, 40, 41), and where the even member is a square is (63, 16, 65). Further examples can be found in appendix C. Interestingly, though, is the fact that there are no Pythagorean triples where *both* first two members are perfect squares.[10] Moreover, you might also want to convince yourself that *only one* side of a Pythagorean triple can have a perfect square.

Figure 4-18 shows a few examples of Pythagorean triples where the smallest member is a perfect square and where the smallest member is a perfect cube.

| **Primitive Pythagorean Triples with the Smallest Member a Perfect Square** | **Primitive Pythagorean Triples with the Smallest Member a Perfect Cube** |
|---|---|
| (9, 40, 41) | (27, 364, 365) |
| (16, 63, 65) | (64, 1023, 1025) |
| (25, 312, 313) | (125, 7812, 7813) |
| (36, 77, 85) | (216, 713, 745)<br>(216, 11663, 11665) |
| (49, 1200, 1201) | (343, 58824, 58825) |
| (64, 1023, 1025) | (512, 65535, 65537) |
| (81, 3280, 3281) | (729, 265720, 265721) |
| (100, 621, 629) | (1000, 15609, 15641)<br>(1000, 249999, 250001) |
| (121, 7320, 7321) | (1331, 885780, 885781) |

**Figure 4-18**

You should be able to see a pattern between the two larger members of each of these Pythagorean triples.

*Curiosity 6*

As you might have expected with the infinitely many Pythagorean triples that exist, there are some that have an area— $\frac{1}{2}ab = mn(m^2 - n^2)$, using Euclid's formula—that is represented by a number that uses nine different digits. For $m = 149$ and $n = 58$, we get

$$a = m^2 - n^2 = 149^2 - 58^2 = 17{,}284$$
$$b = 2mn = 2(149)(58) = 18{,}837$$
$$c = m^2 + n^2 = 149^2 + 58^2 = 653{,}569{,}225$$

10. This was proved by Pierre de Fermat.

This gives us the Pythagorean triple (17,284; 18,837; 653,569,225), whose area is $\frac{1}{2}(17,284)(18,837) = 162,789,354$, where nine nonzero different digits are represented exactly once each. Another such Pythagorean triple is one where $m = 224$ and $n = 153$. Here the Pythagorean triple we get is (26,767; 68,544; 73,585), which represents a right triangle having an area of $\frac{1}{2}(26,767)(68,544) = 917,358,624$. Again all nine nonzero digits are used to express the area number.

To obtain Pythagorean triples representing right triangles whose area is represented by a number having all ten digits used exactly once, we can use the Euclidean formula with $m = 666$ and $n = 5$. This Pythagorean triple (443,531; 6,660; 443,581) has an area represented by the number 1,476,958,230, which uses each of our ten digits exactly once. We can get another such Pythagorean triple using the values of $m = 406$ and $n = 279$. Here we get a Pythagorean triple whose right triangle area is 9,854,271,630. You ought to be able to find the sides of this right triangle.

*Curiosity 7*

Another property of Pythagorean triples $(a, b, c)$ is that they all have the following relationship:

$$\frac{(c-a)(c-b)}{2} \text{ is always a perfect square}$$[11]

11. The converse of this statement is *not* true, namely, that if $\frac{(c-a)(c-b)}{2}$ is a perfect square, then the triple $(a, b, c)$ is a Pythagorean triple. A case in point is the triple (6, 12, 18), where this relationship holds true, but it is not a Pythagorean triple.

Let's try this relationship on a few Pythagorean triples.

For the Pythagorean triple (3, 4, 5) we find

$$\frac{(5-3)(5-4)}{2}=\frac{2}{2}=1$$

which is a perfect square.

For the Pythagorean triple (8, 15, 17) we find

$$\frac{(17-8)(17-15)}{2}=\frac{9\cdot 2}{2}=9$$

which is a perfect square.

For the Pythagorean triple (7, 24, 25) we find

$$\frac{(25-7)(25-24)}{2}=\frac{18\cdot 1}{2}=9$$

which is a perfect square.

If this doesn't convince you that this should be true for all Pythagorean triples, then the following algebraic justification should do the trick. Using our trusty Euclidean formula again: $a=m^2-n^2,\quad b=2mn,$ and $c=m^2+n^2$, we get

$$\frac{(c-a)(c-b)}{2}=\frac{\left(\left(m^2+n^2\right)-\left(m^2-n^2\right)\right)\left(\left(m^2+n^2\right)-2mn\right)}{2}$$

$$=\frac{\left(2n^2\right)(m-n)^2}{2}$$

$$=\left(n^2\right)(m-n)^2$$

which is a perfect square.

*Curiosity 8*

Another curiosity embedded in the properties of Pythagorean triples is that if we consider the Pythagorean triple (5, 12, 13) and place the digit 1 before each member of the triple, then we get (15, 112, 113), which is also a Pythagorean triple. This is believed to be the only case in which a single digit can be placed at the left of each member of a Pythagorean triple to generate another Pythagorean triple.

*Curiosity 9*

Some symmetric Pythagorean triples are also worth highlighting. One is where the second and third members are reverses of one another, and the first member is a palindromic number.[12] Here are two such triples: (33, 56, 65) and (3333, 5656, 6565). Can you find other such "symmetric" pairs of Pythagorean triples?

There are also Pythagorean triples where the first two members are reverses of one another, such as (88209, 90288, 126225). Are there more such triples?

*Curiosity 10*

Naturally we can create palindromic Pythagorean triples by multiplying each of the members of the (3, 4, 5) triple by 11, 111, 1111, ... or by 101, 1001, 10001, ... and so on. We would get Pythagorean triples that will look like this: (33, 44, 55), (333, 444, 555), ..., or (303, 404, 505), (3003, 4004, 5005), ....

On the other hand, there are some Pythagorean triples that contain a few palindromic numbers. Some of these are (20, 99, 101), (252, 275, 373), and (363, 484, 605). The latter example, where the first two

12. A *palindromic number* is one that can be read the same in either direction, such as 13531.

members are palindromes, has many other examples as you can see from figure 4-19.

| **Pythagorean triples with a Pair of Palindromic Numbers** | | |
|---:|---:|---:|
| 3 | 4 | 5 |
| 6 | 8 | 10 |
| 363 | 484 | 605 |
| 464 | 777 | 905 |
| 3993 | 6776 | 7865 |
| 6776 | 23232 | 24200 |
| 313 | 48984 | 48985 |
| 8228 | 69696 | 70180 |
| 30603 | 40804 | 51005 |
| 34743 | 42824 | 55145 |
| 29192 | 60006 | 66730 |
| 25652 | 55755 | 61373 |
| 52625 | 80808 | 96433 |
| 36663 | 616616 | 617705 |
| 48984 | 886688 | 888040 |
| 575575 | 2152512 | 2228137 |
| 6336 | 2509052 | 2509060 |
| 2327232 | 4728274 | 5269970 |
| 3006003 | 4008004 | 5010005 |
| 3458543 | 4228224 | 5462545 |
| 80308 | 5578755 | 5579333 |
| 2532352 | 5853585 | 6377873 |
| 5679765 | 23711732 | 24382493 |
| 4454544 | 29055092 | 29394580 |
| 677776 | 237282732 | 237283700 |
| 27280108272 | 55873637855 | 62177710753 |
| 300060003 | 400080004 | 500100005 |
| 304070403 | 402080204 | 504110405 |
| 276626672 | 458515854 | 535498930 |
| 341484143 | 420282024 | 541524145 |
| 345696543 | 422282224 | 545736545 |
| 359575953 | 401141104 | 538710545 |
| 277373772 | 694808496 | 748127700 |
| 635191536 | 2566776652 | 2644203220 |
| 6521771256 | 29986068992 | 30687095560 |
| 21757175712 | 48337273384 | 53008175720 |
| 30000600003 | 40000800004 | 50001000005 |
| 30441814403 | 40220802204 | 50442214405 |
| 34104840143 | 42002820024 | 54105240145 |

**Figure 4-19**

*Curiosity 11*

Any pair of Pythagorean triples, $(a, b, c)$ and $(p, q, r)$, are related with the following equation: $(c+r)^2-(a+p)^2-(b+q)^2=X^2$, where $X$ is an integer. Let's try to apply this relationship to two Pythagorean triples. We will select the following two Pythagorean triples: (7, 24, 25) and (15, 8, 17).

Applying this relationship we get

$$(25+17)^2-(24+8)^2-(7+15)^2=42^2-32^2-22^2$$
$$=1{,}764-1{,}024-484=256=16^2$$

You might want to try to see if this relationship holds for other pairs of Pythagorean triples. It will.

*Curiosity 12*

Some Pythagorean curiosities are just that: nothing but an unusual number that raises eyebrows. One such curiosity is the Pythagorean triple (693; 1,924; 2,045), which just happens to have an area of 666,666. Readers involved in numerology will recognize this as a sort of double 666, often referred to as the "Number of the Beast," as described in the book of Revelation (13:17–18) in the New Testament of the Christian Bible.

*Curiosity 13*

Consider the first 9 digits of pi's value[13] in groups of three: 314 159 265. The second two groups of three, together with 212, form a Pythagorean triple (159, 212, 265). This newly introduced number, 212, together with 666, forms a quotient that gives a nice approximation of $\pi$. That is, $\frac{666}{212}=3.14\overline{1509433962264}$.

[13] Recall that the value of the ratio of the circumference of a circle to its diameter is $\pi \approx 3.14159265358979 \ldots$.

*Curiosity 14*

In high school algebra we study another type of number called an imaginary number or, in the more general form, a complex number, which is composed of a real and an imaginary component in the form $a + bi$.

Here $a$ is the real part and $b$ is the imaginary part, where $i = \sqrt{-1}$. We can then get powers of $i$ as follows:

$$i^2 = -1, \quad i^3 = -i, \quad i^4 = +1, \quad i^5 = i, \quad i^6 = -1, \quad i^7 = -i, \quad i^8 = +1, \ \ldots$$

Now that we have refreshed our understanding of the nature of a complex number, we can apply it to the Pythagorean triples.

Curiously, we can establish a completely unexpected connection between complex numbers and the Pythagorean triples. That is, the square of a complex number will always yield two legs of a Pythagorean triple. Let's see how this works. We can begin by considering the complex number $3 + 2i$. Squaring this number:

$$(3+2i)^2 = 3^2 + 12i + 4i^2 = 9 + 12i - 4 = 5 + 12i$$

Now, notice that 5 and 12 are the legs of the Pythagorean triple (5, 12, 13). We shall apply this technique once more. This time we will begin with the complex number $5 - 6i$. Squaring this complex number gives us

$$(5-6i)^2 = 5^2 - 60i + 36i^2 = 25 - 60i - 36 = -11 - 60i$$

which indicates the legs of the Pythagorean triple (11, 60, 61).

You may be wondering why this always works, since the complex numbers on the surface would seem to have nothing to do with the Pythagorean triples. We can shed light on this enigma by considering the general case, $a + bi$. We can square this number as follows: $(a + bi)^2 = (a^2 - b^2) + 2abi$. This should look somewhat familiar by now. Yes, it is in the same form as the Euclidean formula for generating Pythagorean triples.

We recall that $\left(a^2 - b^2\right)^2 + \left(2ab\right)^2 = \left(a^2 + b^2\right)^2$ is the Euclidean formula, which has its legs taken from the square of the complex number.

*Curiosity 15*

As our final curiosity of the Pythagorean triples, we find that there is even a systematic way of determining the smallest number that can represent a member of a Pythagorean triple a given number of times, as either a leg or the hypotenuse. For example, the smallest number that can be the side of a Pythagorean triangle exactly once is the number 3. The number 4 can also be a member of only one Pythagorean triple, but it is not the smallest number—3 is the smallest. The number 5 is the smallest number that can be a member of exactly two Pythagorean triples: once as a leg, as in (5, 12, 13), and once as the hypotenuse, as in (3, 4, 5). The smallest number that can be used in exactly three different Pythagorean triples is the number 16. It is used in the following three Pythagorean triples: (12, 16, 20), (30, 16, 34), and (63, 16, 65). Notice that it could not be the length of the hypotenuse. In order for a number to be able to be the hypotenuse, it must have at least one prime factor of the form $4n + 1$. The number 5 is a prime number in this form ($4 \cdot 1 + 1$), while the number 16 does not have a prime factor of this form.[14] Figure 4-20 lists the smallest natural number that can be a member of a Pythagorean triple a given number of times. You will see that the number 40 is the smallest number that can be a member of exactly eight Pythagorean triples. In seven of these eight Pythagorean triples, the number 40 will be a leg and in one of them it will be the hypotenuse—that being (24, 32, 40).

---

[14] The prime factors of 16 are $2 \cdot 2 \cdot 2 \cdot 2$.

## List of Numbers ($n$) That Can Be Used to Represent the Side of a Right Triangle a Specific Number of Times

| Number of Times *n* Can Be the Side of a Right Triangle | *n* | Number of Times *n* Is a Leg of a Right Triangle | Number of Times *n* Is a Hypotenuse of a Right Triangle |
|---|---|---|---|
| 1 | 3 | 1 | 0 |
| 2 | 5 | 1 | 1 |
| 3 | 16 | 3 | 0 |
| 4 | 12 | 4 | 0 |
| 5 | 15 | 4 | 1 |
| 6 | 125 | 3 | 3 |
| 7 | 24 | 7 | 0 |
| 8 | 40 | 7 | 1 |
| 9 | 75 | 7 | 2 |
| 10 | 48 | 10 | 0 |
| 11 | 80 | 10 | 1 |
| 12 | 72 | 12 | 0 |
| 13 | 84 | 13 | 0 |
| 14 | 60 | 13 | 1 |
| 15 | 32768 | 15 | 0 |
| 16 | 192 | 16 | 0 |
| 17 | 144 | 17 | 0 |
| 18 | 524288 | 18 | 0 |
| 19 | 384 | 19 | 0 |
| 20 | 640 | 19 | 1 |
| 21 | 9375 | 16 | 5 |
| 22 | 168 | 22 | 0 |
| 23 | 120 | 22 | 1 |
| 24 | 300 | 22 | 2 |
| 25 | 1536 | 25 | 0 |
| 26 | 520 | 22 | 4 |
| 27 | 576 | 27 | 0 |
| 28 | 3072 | 28 | 0 |
| 29 | 975 | 22 | 7 |
| 30 | 2147483648 | 30 | 0 |
| 31 | 336 | 31 | 0 |
| 32 | 240 | 31 | 1 |
| 33 | 1171875 | 25 | 8 |
| 34 | 1500 | 31 | 3 |
| 35 | 1040 | 31 | 4 |
| 36 | 137438953472 | 36 | 0 |
| 37 | 504 | 37 | 0 |
| 38 | 360 | 37 | 1 |
| 39 | 600 | 37 | 2 |
| 40 | 924 | 40 | 0 |
| 41 | 420 | 40 | 1 |
| 42 | 4608 | 42 | 0 |
| 43 | 50000 | 38 | 5 |
| 44 | 780 | 40 | 4 |
| 45 | 3456 | 45 | 0 |
| 46 | 196608 | 46 | 0 |
| 47 | 9216 | 47 | 0 |
| 48 | 16000 | 45 | 3 |

| 49 | 1344 | 49 | 0 |
|---|---|---|---|
| 50 | 960 | 49 | 1 |
| 51 | 250000 | 45 | 6 |
| 52 | 1008 | 52 | 0 |
| 53 | 720 | 52 | 1 |
| 54 | 1200 | 52 | 2 |
| 55 | 3000 | 52 | 3 |
| 56 | 2621440 | 55 | 1 |
| 57 | 36864 | 57 | 0 |
| 58 | 2688 | 58 | 0 |
| 59 | 1920 | 58 | 1 |
| 60 | 15552 | 60 | 0 |
| 61 | 6291456 | 61 | 0 |
| 62 | 3528 | 62 | 0 |
| 63 | 18446744073709551616 | 63 | 0 |
| 64 | 1800 | 62 | 2 |
| 65 | 121875 | 49 | 16 |
| 66 | 27648 | 66 | 0 |
| 67 | 1848 | 67 | 0 |
| 68 | 840 | 67 | 1 |
| 69 | 2100 | 67 | 2 |
| 70 | 50331648 | 70 | 0 |
| 71 | 1560 | 67 | 4 |
| 72 | 294912 | 72 | 0 |
| 73 | 3024 | 73 | 0 |
| 74 | 2160 | 73 | 1 |
| 75 | 31250000 | 66 | 9 |
| 76 | 6000 | 73 | 3 |
| 77 | 7680 | 76 | 1 |
| 78 | 604462909807314587353088 | 78 | 0 |
| 79 | 1638400 | 77 | 2 |
| 80 | 8840 | 67 | 13 |
| 81 | 286102294921875 | 61 | 20 |
| 82 | 4032 | 82 | 0 |
| 83 | 2880 | 82 | 1 |
| 84 | 4800 | 82 | 2 |
| 85 | 21504 | 85 | 0 |
| 86 | 15360 | 85 | 1 |
| 87 | 7056 | 87 | 0 |
| 88 | 3221225472 | 88 | 0 |
| 89 | 3600 | 87 | 2 |
| 90 | 9000 | 87 | 3 |
| 91 | 781250000 | 80 | 11 |
| 92 | 4718592 | 92 | 0 |
| 93 | 124416 | 93 | 0 |
| 94 | 3696 | 94 | 0 |
| 95 | 1680 | 94 | 1 |
| 96 | 158456325028528675187087900672 | 96 | 0 |
| 97 | 8064 | 97 | 0 |
| 98 | 3120 | 94 | 4 |
| 99 | 9600 | 97 | 2 |
| 100 | 51539607552 | 100 | 0 |

**Figure 4-20**

## The Inradius of a Pythagorean Triangle

Another number that nicely relates to the Pythagorean triple is the length of the inradius (i.e., the radius of the inscribed circle, which is the circle inscribed in a triangle, tangent to the three sides of the triangle) of a Pythagorean triangle. In figure 4-21 the drawn inradius, $r$, is perpendicular to each of the three sides of the right triangle.

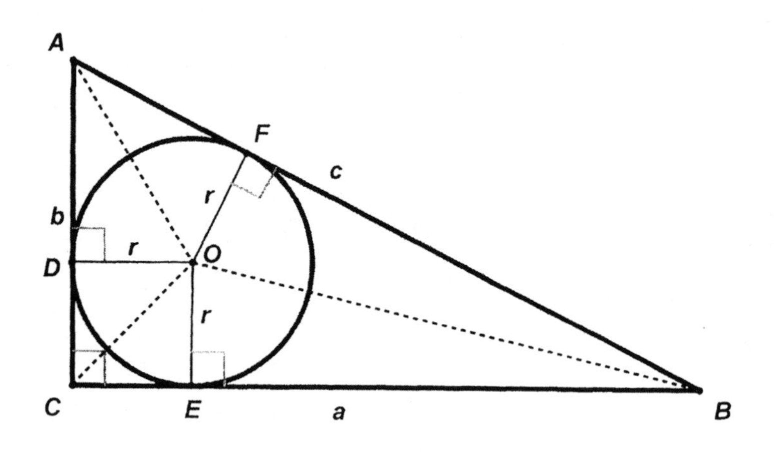


**Figure 4-21**

We can find the area of the right triangle $ABC$ by taking the sum of the areas of $\Delta BOC$, $\Delta AOC$, and $\Delta AOB$ as follows:

$$Area\ \Delta ABC = \frac{ra}{2} + \frac{rb}{2} + \frac{rc}{2} = r\left(\frac{a+b+c}{2}\right) = rs$$

where $s$ is the semiperimeter (i.e., half the perimeter of the triangle $ABC$).

We can then represent $r$ as $\frac{A}{s}$ or $\frac{2 \times \text{area}}{\text{perimeter}}$. We know that the area of the right triangle is $\frac{ab}{2}$. Therefore, $r = \frac{ab}{a+b+c}$.

We once again turn to the Euclidean formula for finding primitive Pythagorean triples ($a = m^2 - n^2$, $b = 2mn$, and $c = m^2 + n^2$). Substituting these values for $a$, $b$, and $c$ we get the following:

$$
\begin{aligned}
r &= \frac{(m^2 - n^2)(2mn)}{(m^2 - n^2) + (2mn) + (m^2 + n^2)} \\
&= \frac{(m^2 - n^2)(2mn)}{2m^2 + 2mn} \\
&= \frac{(m - n)(m + n)2mn}{2m(m + n)} \\
&= (n)(m - n)
\end{aligned}
$$

This formula[15] for $r$ tells us that the inradius is always an integer, since $m$ and $n$ are integers. If you inspect the list of primitive Pythagorean triples in appendix C, you will notice that for every positive integer $r$, there is a primitive Pythagorean triple associated with it. There is at least one nonprimitive Pythagorean triple associated with each of the integer values of $r > 1$.

As we continue to search for patterns among the Pythagorean triples, we can now include the inradius. If we look at a list (figure 4-22) of the first few odd inradii and their accompanying Pythagorean triples, we ought to be able to see a pattern evolving.

For every prime number inradius $r > 2$, there will be exactly two primitive Pythagorean triples associated with it. In figure 4-22 we list the odd integers $r$, which included two nonprimes (in shaded rows), where there are sometimes more than two Pythagorean triples.[16] For each of the primes listed in figure 4-22, you will also notice a pattern for the types of Pythagorean triples; one triple has the hypotenuse one greater than the larger leg and the other triple has the hypotenuse two

15. This formula can be traced back to the Chinese mathematician Liu Hui (ca. 220–280 CE), which he mentions in writings of the year 263.

16. According to Neville Robbins (*Fibonacci Quarterly* 44 [2006]: 368–69), the number of primitive Pythagorean triples when the inradius $r$ is odd is equal to $2^{(\text{number of distinct prime factors of } r)}$, and when $r$ is even the number of primitive Pythagorean triples is equal to $2^{(\text{number of distinct prime factors of } r)-1}$.

greater than the larger leg. There are many more such patterns that can be found in the list of Pythagorean triples as they relate to the inradius.

| Inradius | Primitive Pythagorean triples |
|---|---|
| 3 | (7, 24, 25), (8, 15, 17) |
| 5 | (11, 60, 61), (12, 35, 37) |
| 7 | (15, 112, 113), (16, 63, 65) |
| 9 | (19, 180, 181), (20, 99, 101) |
| 11 | (23, 264, 265), (24, 143, 145) |
| 13 | (27, 364, 365), (28, 195, 197) |
| 15 | (31, 480, 481), (32, 255, 257), (39, 80, 89), (48, 55, 73) |
| 17 | (35, 612, 613), (36, 323, 325) |
| 19 | (39, 760, 761), (40, 399, 401) |
| 21 | (43, 924, 925), (44, 483, 485), (51, 140, 149), (60, 91, 109) |
| 23 | (47, 1104, 1105), (48, 575, 577) |
| 25 | (51, 1300, 1301), (52, 675, 677) |

**Figure 4-22**

## Beyond the Pythagorean Theorem

It is tempting to want to generalize the Pythagorean Theorem to powers greater than 2. In other words, can we find a triple of integers ($a$, $b$, c) that would satisfy the equation $a^n + b^n = c^n$ for values of $n > 2$? This has been tried, despite the claim by Pierre de Fermat that no such solution exists. His claim was written in the margin of one of his mathematics books and it challenged mathematicians to prove it for 357 years, until 1994, when Dr. Andrew Wiles (1953–) proved that Fermat's Conjecture (or as it has been popularly known, "Fermat's Last Theorem") was actually correct. However, in the span of more than three numbers we can get some interesting analogous relationships.

$$3^3 + 4^3 + 5^3 = 6^3$$

$$30^4 + 120^4 + 272^4 + 315^4 = 353^4$$

$$19^5 + 43^5 + 46^5 + 47^5 + 67^5 = 72^5$$

$$127^7 + 258^7 + 266^7 + 413^7 + 430^7 + 439^7 + 525^7 = 568^7$$

$$90^8 + 223^8 + 478^8 + 524^8 + 748^8 + 1{,}088^8 + 1{,}190^8 + 1{,}324^8 = 1{,}409^8$$

Notice that the number of terms in each of the above sums is the same as the number in the exponent.

Pythagorean triples offer an almost unlimited plethora of opportunities for finding amazing numerical relationships, which the reader is encouraged to pursue.

# Chapter 5

# The Pythagorean Means

Normally, when we refer to the average of a set of numbers, we mean the *arithmetic mean*. That is, the sum of the numbers, whose "average" you seek, divided by the number of such numbers involved. In high school we usually encounter another type of average called the *geometric mean* (or sometimes called the mean proportional). This is the number that is in the "means position" in a proportion, such as $x$ in the following proportion: $\frac{p}{x} = \frac{x}{q}$, where we say that $x$ is the mean proportional between $p$ and $q$. There is yet another mean—one that is rarely encountered—and it is called the *harmonic mean*. This harmonic mean is simply the reciprocal of the average (arithmetic mean) of the reciprocals of the numbers being considered. Together, these three means are often referred to as the "Pythagorean means," since it is assumed that Pythagoras learned about these means during his stay in Mesopotamia and then widely used them. We have learned over the years that the Pythagoreans were very concerned about measuring and comparing quantities, which quite likely gave rise to their popularizing these mean measures. Still today we have good evidence that these means were already known to the Babylonians, who employed them in their calculations.

We begin our investigation of the Pythagorean means with what might be considered the least popular and most often neglected mean, the harmonic mean, first in its role as a useful problem-solving tool. Then, after properly defining it, we will compare it to the other more familiar means. Consider the following problem:

> Mr. Wagner travels from New York City to Washington, DC, a distance of 240 miles at an average speed of 60 miles per hours. On his return trip along the same route, he encounters inclement weather and averages only 30 miles per hours. What is Mr. Wagner's speed for his entire trip?

You might be quite surprised that 45 miles per hours is *not* the correct answer. Surely 45 is the "average" (or arithmetic mean) between 60 and 30. However, rates (of speed, in this case) cannot be treated as simple quantities. Mr. Wagner is driving *twice* as much time at the rate of 30 miles per hour as at 60 miles per hour. Hence, it would be incorrect to give both rates the same "weight." By properly adjusting the weights of the speeds—that is, giving the 30 mph twice the weight as the 60 mph—one can get the correct average speed:

$$\frac{30+30+60}{3}=40$$

Such a simple solution would hardly be expected if one rate were not a multiple of the other as was the case here where 60 was twice 30. In that case, the average rate for the entire trip would be found by obtaining the quotient of the *total distance* traveled (here $2\times 240=480$ miles) divided by the *total time*[1] traveled (here $4+8=12$ hours), which, in this case, is 40 miles per hour.

1. We find the time going 240 miles to Washington at 60 mph will take 4 hours, and the return trip at 30 mph will take twice as long: 8 hours.

Let's now consider a general case—so that we may establish a formula—where we seek the average rate between two rates, $r_1$ and $r_2$, each over a distance $d$.[2]

We find the time for each speed: $t_1 = \frac{d}{r_1}$ and $t_2 = \frac{d}{r_2}$

The total time, $t$, $= t_1 + t_2 = d\left(\frac{1}{r_1} + \frac{1}{r_2}\right) = d\left(\frac{r_1 + r_2}{r_1 r_2}\right)$

However, the average rate $r$ is determined by $\frac{2d}{t}$

$$\text{Thus, } r = \frac{2d}{t} = \frac{2d}{d\left(\frac{r_1 + r_2}{r_1 r_2}\right)} = \frac{2r_1 r_2}{r_1 + r_2}$$

This gives us a formula for finding the average rate (of speed) between two given rates over *equal* distances. An inspection will show that this last expression, $\frac{2r_1r_2}{r_1+r_2}$, is the reciprocal of the arithmetic mean[3] of the reciprocals $r_1$ and $r_2$. This "average" is referred to as the *harmonic mean* between $r_1$ and $r_2$.

Just as arithmetic and geometric means are derived from arithmetic[4] and geometric[5] sequences (or sequences of numbers), re-

---

2. Recall that the relationship among rate, time, and distance is *rate* X *time* = *distance*. Therefore, $r = \frac{d}{t}$.

3. This somewhat complex statement can be seen symbolically as $\frac{1}{\frac{\frac{1}{r_1} + \frac{1}{r_2}}{2}}$, or you can see this as $\frac{1}{\frac{1}{2}\left(\frac{1}{r_1} + \frac{1}{r_2}\right)}$.

4. An *arithmetic sequence* is a progression of numbers with a common difference between them. For example, 2, 5, 8, 11, 14, ... is an arithmetic sequence since the difference between terms is the same, namely, 3.

spectively, so we may now establish a harmonic sequence. From the previous definition of a harmonic mean, we find that a harmonic sequence is one where the reciprocals of its members form an arithmetic sequence. That is, if $a$, $b$, and $c$ are in harmonic sequence, then $\frac{1}{a}$, $\frac{1}{b}$, and $\frac{1}{c}$ are in arithmetic sequence. If we establish that $\frac{1}{b}$ is the arithmetic mean between $\frac{1}{a}$ and $\frac{1}{c}$, then $b$ is the harmonic mean between $a$ and $c$. Most work with a harmonic sequence is done by first converting it to an arithmetic sequence (by taking the reciprocals of each of the terms), but to some extent this limits computations with harmonic sequences. Although general formulas exist for finding the sum of an arithmetic series or a geometric series, no such formula exists for finding the sum of a harmonic series.

The definition, above, of a harmonic mean (HM) may be extended to find the harmonic mean between 3, 4, or $n$ numbers:

For numbers $r$ and $s$

$$\text{HM} = \frac{2}{\frac{1}{r} + \frac{1}{s}} = \frac{2rs}{r+s}$$

For numbers $r$, $s$, and $t$

$$\text{HM} = \frac{3}{\frac{1}{r} + \frac{1}{s} + \frac{1}{t}} = \frac{3rst}{st + rt + rs}$$

---

5. A *geometric sequence* is a progression of numbers with a common multiple between them. For example, 2, 6, 18, 54, ... is a geometric sequence since the multiple between terms is the same, namely, 3.

For numbers $r$, $s$, $t$, and $u$

$$\text{HM} = \frac{4}{\frac{1}{r}+\frac{1}{s}+\frac{1}{t}+\frac{1}{u}}$$
$$= \frac{4rstu}{rst + rtu + stu + rsu}$$

For numbers $r_1, r_2, r_3, \ldots, r_n$

$$\text{HM} = \frac{n}{\frac{1}{r_1}+\frac{1}{r_2}+\frac{1}{r_3}+\ldots+\frac{1}{r_n}}$$
$$= \frac{n(r_1 \cdot r_2 \cdot r_3 \cdot \ldots \cdot \ldots \cdot r_{n-1} \cdot r_n)}{\sum_{k=1}^{n} r_1 \cdot r_2 \cdot r_3 \cdot \ldots \cdot \hat{r}_k \cdot \ldots \cdot r_n}$$

where $\hat{r}_k$ indicates that the $r_k$th factor is omitted.

Part of the beauty of the harmonic mean is that it can be used to find the average of rates of various kinds—as long as the base is the same for each rate. Here are two examples of some of these applications:

1. Lisa bought 2 dollars' worth of each of three different kinds of pencils, priced at 2¢, 4¢, and 5¢ each, respectively. What is the average price she paid per pencil?

2. In July, David got 30 hits for a batting average of .300; however, in August he got 30 hits for a batting average of .400. What is David's batting average for July and August?

In each case we seek the harmonic mean to answer the question, since we seek the average rate—in the first case, rate of purchase, and in the second case, rate of hits—and the "base" for each of the problems remains the same.

Because of the apparent novelty of the harmonic sequence and its applications in projective geometry,[6] it would be interesting to consider a geometric representation of a *harmonic sequence*. In figure 5-1, the points of intersection, $D$ and $E$, of the interior and exterior angle bisectors of $\Delta BAC$ with $BC$ determine a harmonic sequence of $BD$, $BC$, and $BE$. To verify this, consider $\Delta ABC$, where $AD$ bisects $\angle BAC$, $AE$ bisects $\angle CAF$, and $B$, $D$, $C$, and $E$ are collinear.[7]

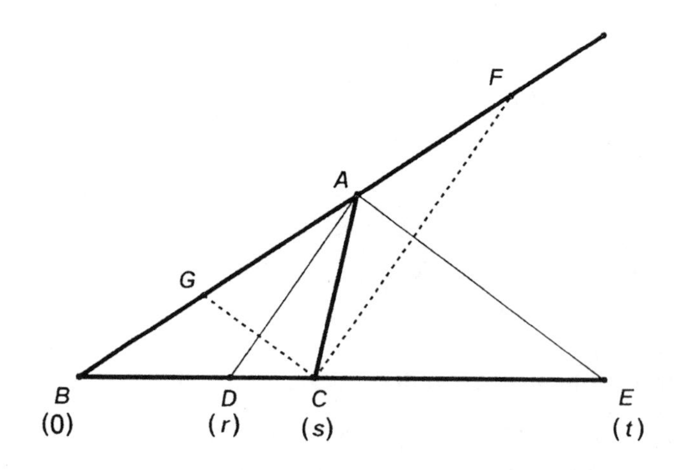


**Figure 5-1**

It can easily be proved that for *exterior* angle bisector $AE$, the following proportion holds true: $\frac{BE}{CE} = \frac{AB}{AC}$.[8] Similarly, for *interior*

---

6. *Projective geometry* is a branch of mathematics that investigates properties of figures that are invariant when projected from a point to a line or plane.

7. Points are *collinear* if they lie on the same straight line.

8. To prove this, we will begin by drawing $GC \parallel AE$. This will give us $\angle EAC = \angle ACG$ and $\angle FAE = \angle AGC$. However, $\angle EAC = \angle FAE$; therefore, $\angle ACG = \angle AGC$. From the isosceles $\Delta AGC$, $AG = AC$. Thus, $\frac{BE}{CE} = \frac{AB}{AG} = \frac{AB}{AC}$.

angle bisector $AD$, we get the proportion $\frac{BD}{CD} = \frac{AB}{AC}$.[9] Therefore, $\frac{BE}{CE} = \frac{BD}{CD}$, or $\frac{CD}{CE} = \frac{BD}{BE}$. It is then said that the points $B$ and $C$ separate the points $D$ and $E$ harmonically.

Now suppose $BDCE$ is a number line with $B$ as the zero point, point $D$ at coordinate $r$, point $C$ at coordinate $s$, and point $E$ at coordinate $t$. This allows us to say that $BD = r$, $BC = s$, and $BE = t$.

We shall show that the sequence $r$, $s$, $t$ is a harmonic sequence.

Since $\frac{CD}{CE} = \frac{BD}{BE}$, $\frac{BC-BD}{BE-BC} = \frac{BD}{BE}$, or $\frac{s-r}{t-s} = \frac{r}{t}$

Therefore, $t(s-r) = r(t-s)$ and $ts - tr = rt - rs$. Dividing each term by $rst$, we get $\frac{1}{r} - \frac{1}{s} = \frac{1}{s} - \frac{1}{t}$, which indicates that the sequence $\frac{1}{t}$, $\frac{1}{s}$, $\frac{1}{r}$ forms an arithmetic sequence, since there is a common difference between terms. Then the sequence of reciprocals, $r$, $s$, $t$, forms a harmonic sequence.

※※※

The geometric interpretations of these algebraic concepts are quite fascinating and provide a nice visual comparison of the Pythagorean means. Consider the length of the segment parallel to the bases of a trapezoid and containing the point of intersection of its diagonals, and with its endpoints on the sides of the trapezoid. The length of this segment, $FGE$ (figure 5-2), is the harmonic mean between the lengths of the bases $AD$ and $BC$ of trapezoid $ABCD$, where $AD \parallel BC$, $FGE \parallel BC$ ($\parallel$ standing for "is parallel to"), and $F$ and $E$ are points on $AB$ and $CD$, respectively.

9. The proof is done by drawing $CF \parallel AD$; then as done in the previous footnote, $AF = AC$; $\frac{BD}{CD} = \frac{AB}{AF} = \frac{AB}{AC}$.

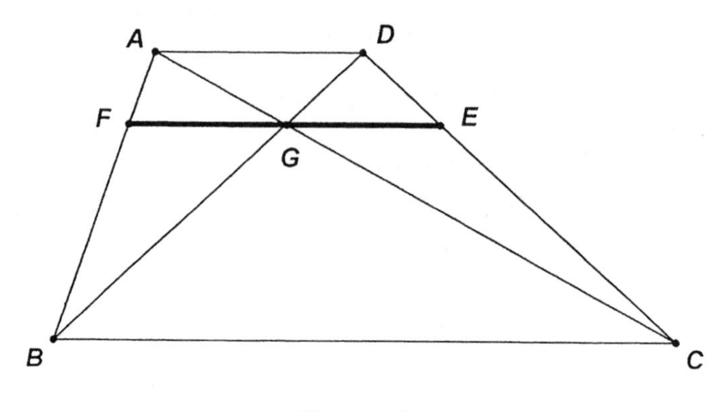


**Figure 5-2**

Using little more than the proportionality of similar triangles, we can show this to be true. We have $GF \parallel BC$, and $\Delta AFG \sim \Delta ABC$ and $\frac{AF}{FG} = \frac{AB}{BC}$. Similarly, since $GF \parallel AD$, we can establish that $\Delta GBF \sim \Delta DBA$ and $\frac{BF}{FG} = \frac{AB}{AD}$. Therefore, by adding these last two equations we get $\frac{AF}{FG} + \frac{BF}{FG} = \frac{AB}{BC} + \frac{AB}{AD}$. However, we notice that $AF + BF = AB$.

Therefore, we can add the fraction on the left side to get

$$\frac{AB}{FG} = \frac{AB}{BC} + \frac{AB}{AD}$$

and then with some algebraic manipulation we get

$$FG = \frac{BC \cdot AD}{BC + AD}$$

In a similar manner, we can get

$$EG = \frac{BC \cdot AD}{BC + AD}$$

Therefore,

$$EF = FG + EG = \frac{2 \cdot BC \cdot AD}{BC + AD}$$

This establishes that $EF$ is the harmonic mean between $BC$ and $AD$.

This trapezoid provides us with a convenient way to compare the sizes of the Pythagorean means geometrically—and we hope convincingly, as well. For our geometric comparison of these three means we will consider the same trapezoid $ABCD$ (this time shown in figure 5-3).[10]

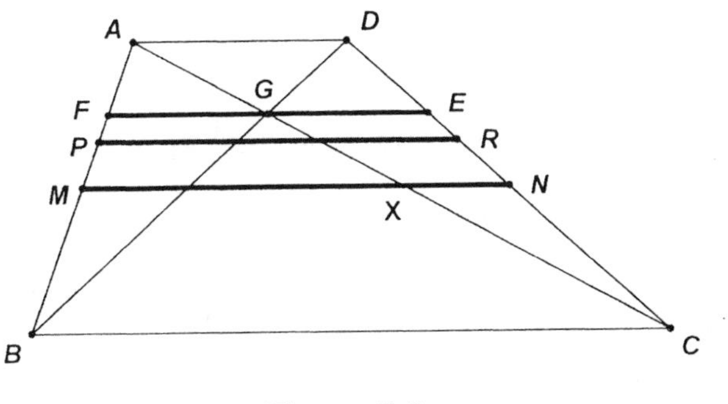


**Figure 5-3**

We will consider the three Pythagorean means in ascending order of magnitude:

- As we have shown in figure 5-2, $FGE$ is the *harmonic mean* between $AD$ and $BC$.
- The length of the line segment $PR$, which is parallel to the bases and having its endpoints in the legs so the trapezoid $ADRP \sim$ trapezoid $PRCB$, is the *geometric mean* between

10. To discover more relationships with this figure, see Howard Eves, "Means Appearing in Geometric Figures," *Mathematics Magazine* 76, no. 4 (October 2003): 292–94.

$AD$ and $BC$. This can be easily shown by comparing the corresponding sides of the two similar trapezoids. That is,

$$\frac{AP}{PB} = \frac{AD}{PR}, \text{ and } \frac{AP}{PB} = \frac{PR}{BC}; \text{ therefore, } \frac{AD}{PR} = \frac{PR}{BC}$$

This shows that $PR$ is the geometric mean between $AD$ and $BC$.

- The length of the line segment $MN$, which joins the midpoints of the legs of the trapezoid $ABCD$ (called the median of the trapezoid), is the *arithmetic mean* between $AD$ and $BC$. This can be demonstrated by considering the similar triangles $AMX$ and $ABC$. This gives us the proportion $\frac{MX}{BC} = \frac{AM}{AB} = \frac{1}{2}$, or $MX = \frac{1}{2}BC$. Similarly, using similar triangles $CNX$ and $CDA$, we can establish that $NX = \frac{1}{2}AD$.

  Therefore, $MN = MX + NX = \frac{1}{2}BC + \frac{1}{2}AD = \frac{BC+AD}{2}$, which shows that $MN$ is the arithmetic mean between $BC$ and $AD$.

In figure 5-3 we can clearly see that the lengths of $MN > PR > FE$, which establishes, geometrically, that arithmetic mean (AM) > geometric mean (GM) > harmonic mean (HM). These means would be equal to each other if the trapezoid $ABCD$ were to be transformed into a parallelogram.

To get a rather nice comparison of the relative magnitudes of the Pythagorean means—the arithmetic, geometric and harmonic means—we will use the diagram in figure 5-4.[11]

11. This demonstration is attributed to Pappus of Alexandria (ca. 320 CE).

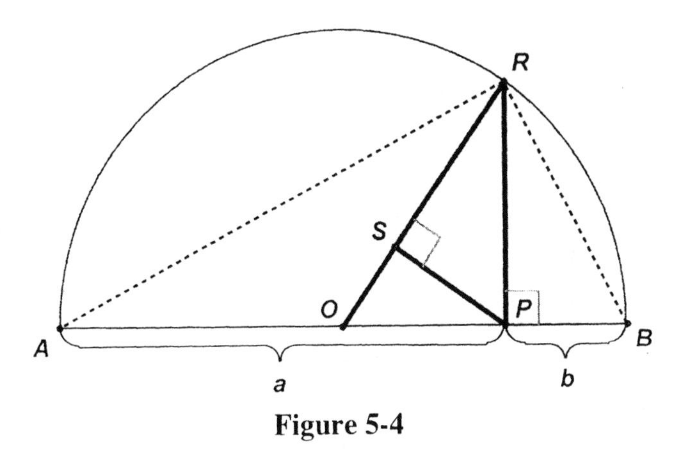


**Figure 5-4**

*To locate the arithmetic mean in figure 5-4:*

Consider the semicircle in figure 5-4. The diameter is $AOPB$ with $AO = OB$ and $PR \perp APB$. Also $PS \perp RSO$. Let $AP = a$ and $PB = b$. Since $RO = \frac{1}{2}AB = \frac{1}{2}(AP + PB) = \frac{1}{2}(a+b)$, $RO$, the radius of the semicircle, is the *arithmetic mean* (AM) between $a$ and $b$.

*To locate the geometric mean in figure 5-4:*

Consider right triangle $ARB$. Since $\Delta BPR \sim \Delta RPA$, $\frac{PB}{PR} = \frac{PR}{AP}$, or $PR^2 = AP \cdot PB = ab$. Therefore, $PR = \sqrt{ab}$. Thus $PR$, the altitude on the hypotenuse of right triangle $ARB$, is the *geometric mean* (GM) between $a$ and $b$.

*To locate the harmonic mean in figure 5-4:*

Consider right triangles $RPO$ and $RSP$, where $PS$ is the altitude on the hypotenuse $OR$.

Since $\Delta RPO \sim \Delta RSP$, $\frac{RO}{PR} = \frac{PR}{RS}$.

Therefore, $RS = \frac{PR^2}{RO}$.

But $(PR)^2 = ab$ and $RO = \frac{1}{2}AB = \frac{1}{2}(a+b)$.

Thus, $RS = \frac{ab}{\frac{1}{2}(a+b)} = \frac{2ab}{a+b}$, which is the *harmonic mean* (HM) between $a$ and $b$.

*Now for the comparisons of mean sizes:*

The hypotenuse of a right triangle is its longest side, and so:

- In right $\Delta ROP$, hypotenuse $RO$ is greater than leg $PR$.
- In right $\Delta RSP$, hypotenuse $PR$ is greater than leg $RS$.

Therefore, $RO > PR > RS$. However, since it is possible for these triangles to degenerate, that is, where all these lines coincide—when $RO \perp AB$ (or $a = b$)—we can extend the comparison to the following: $RO \geq PR \geq RS$. Therefore, we have established geometrically that $\text{AM} \geq \text{GM} \geq \text{HM}$.

To further "position" the geometric mean between the arithmetic and harmonic means, recall that $\frac{RS}{PR} = \frac{PR}{RO}$. Therefore,

$$PR^2 = RO \cdot RS\text{, or } PR = \sqrt{RO \cdot RS}$$

In other words, $\text{GM} = \sqrt{(\text{AM})(\text{HM})}$, or "the *geometric mean* is the geometric mean between the arithmetic mean and the harmonic mean."

Justifying algebraically the comparison of the arithmetic and geometric means using two "numbers" can be done very simply.[12]

---

12. An algebraic comparison, using more than two numbers, of these three means is also quite interesting. But to maintain the flow of the discussion of these fascinating (and related) Pythagorean means, we refer the motivated reader to appendix B.

We begin with a common fact about the two positive numbers $a$ and $b$: $(a-b)^2 \geq 0$, which can be written as $a^2 - 2ab + b^2 \geq 0$.

By adding $4ab$ to both sides of the inequality, we get

$$a^2 + 2ab + b^2 \geq 4ab$$

$$(a+b)^2 \geq 4ab$$

$$\frac{(a+b)^2}{4} \geq ab$$

Taking the positive square root yields $\frac{a+b}{2} \geq \sqrt{ab}$, which tells us that AM $\geq$ GM.[13]

Let us use these convenient two positive numbers, $a$ and $b$, to show the comparison of the geometric and harmonic means. From the relationship we established above, $a^2 + 2ab + b^2 \geq 4ab$, we can multiply both sides by $ab$ to get the following:

$$ab(a+b)^2 \geq (4ab)(ab)$$

Therefore, $ab \geq \frac{4a^2b^2}{(a+b)^2}$, or $\sqrt{ab} \geq \frac{2ab}{a+b}$, which tells us that GM $\geq$ HM.[14]

The comparisons of the relative magnitudes of these Pythagorean means should deepen your insight into them. There are some other striking comparisons and unexpected properties of these Pythagorean means that should give you an enhanced picture of the beauty of the mathematics they provide.

---

13. *Note*: if $a \neq b$, then AM > GM, but if $a = b$, then AM = GM.

14. Note that if $a \neq b$, then GM > HM, but if $a = b$, then AM = GM.

The following is a collection of such "nuggets." Their justifications that use nothing beyond early high school mathematics are provided in appendix B.

*Nugget 1:*

In figure 5-5, $\overarc{APB}$ is a semicircle and $PT$ is tangent to circle $O$ at $P$. $PC \perp ABT$ at $C$.

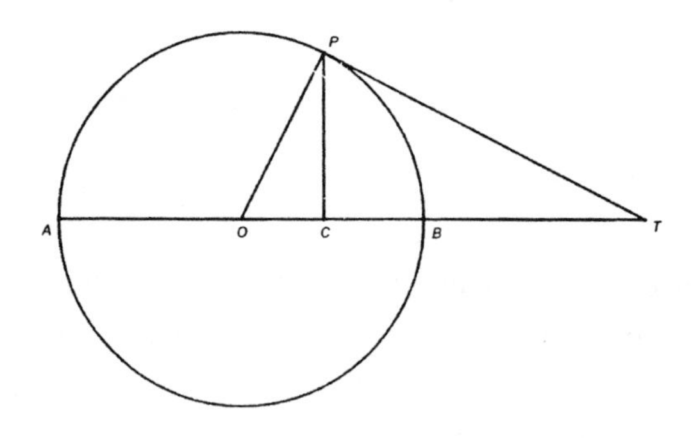


**Figure 5-5**

We can show (using the diagram in figure 5-5) that the following are true:

$TO$ is the arithmetic mean between $AT$ and $BT$.

$PT$ is the geometric mean between $AT$ and $BT$.

$TC$ is the harmonic mean between $AT$ and $BT$.

Then we can justify that $TO \geq PT \geq TC$.

*Nugget 2:*

Consider a given rectangle and a square with the following variations:

- If the rectangle and the square have the *same perimeter*, then the side of the square is the *arithmetic mean* between the sides of the rectangle.
- If the rectangle and the square have the *same area*, then the side of the square is the *geometric mean* between the sides of the rectangle.
- If the rectangle and the square have the *same ratio of area to perimeter*, then the side of the square is the *harmonic mean* between the sides of the rectangle.

*Nugget 3:*

The number of vertices of a cube is the harmonic mean between the number of its edges and the number of its faces.

*Nugget 4:*

In figure 5-6, $P$ is a point on $AB$ of $\Delta ABC$ such that $MP \cong NP$ and $MP \parallel CNB$ and $NP \parallel AMC$. We can conclude that $MP + NP$ is the harmonic mean between $AC$ and $BC$.

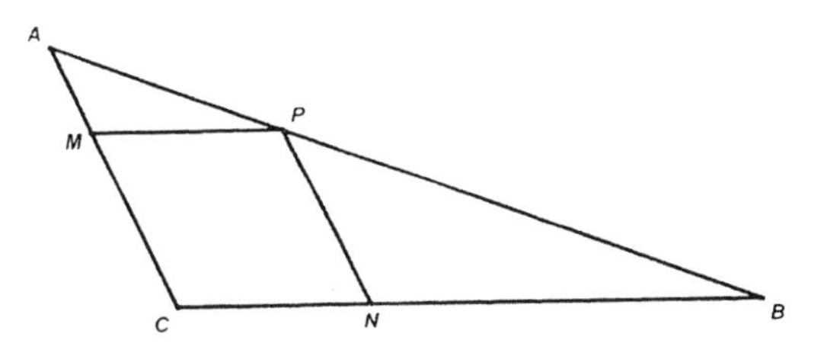


**Figure 5-6**

*Nugget 5:*

In figure 5-7, square $PQRS$ has vertices $P$ and $Q$ on $AC$ and $AB$, respectively, and vertices $S$ and $R$ on $BC$. Here the semiperimeter of the square is the harmonic mean between the lengths of altitude *AH* and base *BC*.

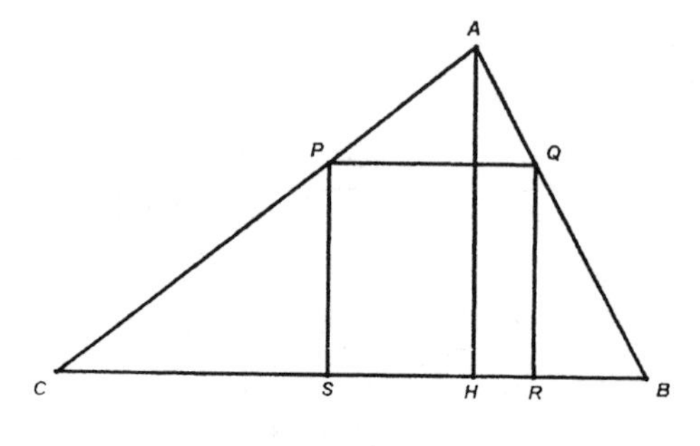


**Figure 5-7**

Now that you have an understanding of the three Pythagorean means, it might be enriching to note that there are other means. These other types of means include the following:[15]

The *Heronian mean* between $a$ and $b$: $\dfrac{a+b+\sqrt{ab}}{3}$

The *contra harmonic mean* between $a$ and $b$: $\dfrac{a^2+b^2}{a+b}$

The *quadratic mean*[16] between $a$ and $b$: $\sqrt{\dfrac{a^2+b^2}{2}}$

The *centroidal mean* between $a$ and $b$: $\dfrac{2\left(a^2+ab+b^2\right)}{3(a+b)}$

15. For more information on these means, see Howard Eves, "Means Appearing in Geometric Figures," *Mathematics Magazine* 76, no. 4 (October 2003): 292–94.

16. The quadratic mean is sometimes referred to as the root-mean square.

# Chapter 6

# Tuning the Soul: Pythagoras and Music[1]

What constitutes a musical note? Is any pitched sound a note? This would seem to be the commonsense answer but it hardly withstands scrutiny. First of all, at different times in Western music history, the same pitch has been designated by a different letter name (i.e., orchestras during the Baroque period tuned well below today's standard of A = 440 cycles per second). Moreover, even within a single tuning system, the same pitch can serve as different notes in different contexts. As a sort of experiment to illustrate this fact, sing the melody "Happy Birthday." Now, starting on the same pitch with which you began "Happy Birthday," sing "Frère Jacques." Both tunes are illustrated in Music Example 1. However, for this experiment, it is not necessary that you read musical notation (as long as you know both tunes), nor is it necessary that you sing these tunes in the keys in which we have notated them (as long as you begin both songs on the same pitch).

---

1. This chapter was contributed by Dr. Chadwick Jenkins, assistant professor of music at the City College of the City University of New York.

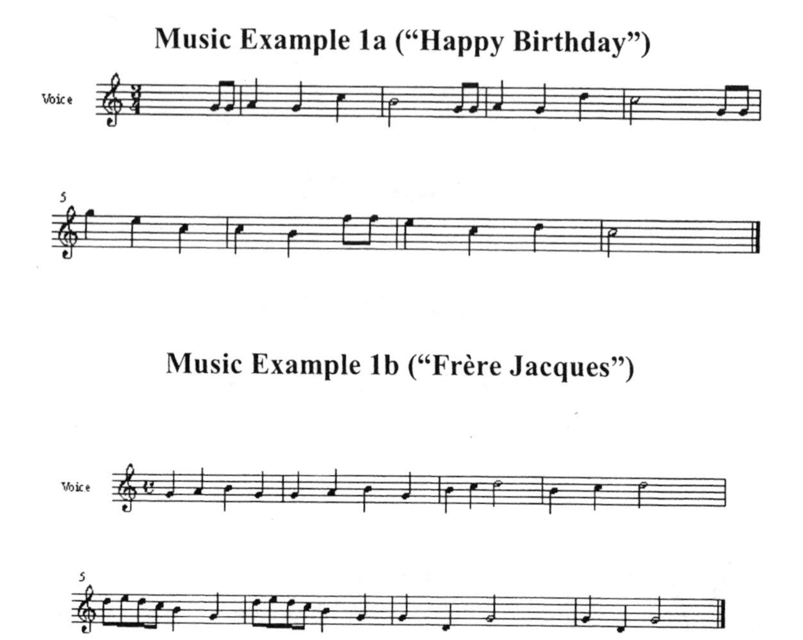


**Music Example 1a ("Happy Birthday")**

**Music Example 1b ("Frère Jacques")**

Notice that in our transcriptions both songs begin with the pitch G (notated on the second line up from the bottom of the staff). However, "Happy Birthday" ends on C (notated on the third space up from the bottom of the staff) while "Frère Jacques" ends on the G with which it began. Yet at the conclusion of both melodies you do not feel as though the given song is incomplete. That is to say, both songs end on the tonic note of the key (the tonic is the *note* that gives a piece a sense of conclusion). The tonic note for "Happy Birthday" is C in this case, and for "Frère Jacques" the tonic note is G. That means that one song ("Frère Jacques") begins with the tonic note and the other song ("Happy Birthday") begins with a nontonic note and yet both began with the pitch G. This should convince you that a musical note (which is defined by its function within an array of other notes) is not commensurate with musical pitch even though any given musical note will necessarily manifest itself *as* a pitch.

This observation, however, implies that in some sense the relationships among notes must be prior to any given note per se (that is, taken on its own). We know this to be true whenever we listen to a less-than-talented amateur singer perform—say, on *American Idol*. Even if the contestant performs a song we have never heard (and therefore cannot know how it "ought" to go), we recognize when our hapless singer falters with respect to pitch. We then say that the contestant is singing "out of tune," by which we mean that the notes in relationship to each other do not lie in what we consider to be a tuned scale.

The significant difference between pitches and notes gives rise to a host of philosophical problems for music theory. Most important, how do we account for the relationships that allow pitches to function properly as notes? That is, do we simply measure the distances between pitches or do we seek some measurable property of the pitches themselves that allows us to determine their relationships to other pitches, thereby constituting a system of notes? As it turns out, this may be the oldest question in the history of music theory. Moreover, Pythagoras's answer to that question not only served as the foundation for a considerable amount of music theoretical speculation over the ensuing centuries but also seems to have launched one of the earliest and most influential schools of philosophy within the history of Western thought.

## Pythagoras the Man/Pythagoras the Legend

Of course, we have no direct knowledge of Pythagoras. He left no writings behind, and as in any enduring school of thought, his followers undoubtedly altered and expanded Pythagoras's ideas after his death. These changes allow a system of thought to remain vibrant and respond to the concerns of later periods, but in this case they make it nearly impossible for a historian to discern what Pythagoras's contributions actually were, and what was added later. Although he was born in the middle of the sixth century BCE, most of the earliest writings that describe his philosophy derive from the late fifth and early fourth centuries BCE. To make matters worse, the difficulty in ascertaining Pythagoras's precise teachings is compounded by the fact that he was the founder and head of a rather secretive cult—so secretive indeed that his followers preferred death to revealing the mysterious Pythagorean prohibition on the eating of beans![2] Our best sources for the thought of Pythagoras and the early Pythagoreans are of three types: (1) writers such as Philolaus, who seem to have been relatively strict Pythagoreans (albeit with concerns and emphases of their own); (2) the writings of Plato, who was clearly impressed with the achievements of the Pythagoreans and adopted many of their ideas within his own philosophical system; and (3) the writings of Aristotle, who wrote an entire book on the Pythagoreans (now lost) and articulated and criticized their ideas in many of his other works.

However, the lack of direct knowledge of Pythagoras's teachings did not prevent ideas associated with this early philosopher from proving most influential. Indeed, it may have been the shadowy, quasi-mythical nature of Pythagoras that made him such a mainstay of music-theoretical thought from antiquity to the dawn of the modern age. The mythic Pythagoras served as a potent symbol of the desire to discover the hidden secrets of music and music's connections to the organizing factors of the universe as well as to the inner workings of the human soul. Even today, whenever someone uses an electronic tuner to tune his instrument or refers to a perfect as opposed to imperfect consonance (as discussed below), vestiges of Pythagorean musical thought remain active. In what follows, we will not be able simply to describe the music-theoretical ideas of the historical Pythagoras (an impossible project) but rather we will trace the impact of ideas accepted as Pythagorean at various times within the history of musical thought.

---

2. Differing accounts survive for the reason behind the prohibition on eating beans. For a concise summary, see Christoph Riedweg, *Pythagoras: His Life, Teaching, and Influence*, trans. Steven Rendall (Ithaca, NY: Cornell University Press, 2002), pp. 69–71.

## Pythagoras and the Blacksmith's Shop: An Originary Tale of Musical Thought

Pythagoras was taking a stroll one day when he happened to pass a blacksmith's shop in which several blacksmiths were pounding various pieces of metal into shape. He noticed that certain composite sounds, which he recognized as consonant musical intervals, arose as a result of their labors. He decided to investigate. His first hypothesis (depending on the source) was that the sound emanated from the anvils. Therefore, he had the blacksmiths rotate their positions. However, the same musical intervals proceeded from the same pairs of blacksmiths. So he postulated that the sounds were the result of the varying force with which each blacksmith struck the anvil. Therefore, he had the blacksmiths exchange hammers. However, the same musical intervals proceeded from the same pairs of hammers. Hence, Pythagoras decided to weigh the hammers, and he discovered that the proportions between the weights of the hammers gave rise to specific musical intervals.

It is important to realize that Pythagoras's discovery depended upon a prior *musical* understanding of consonance, or harmonious sound. That is to say, Pythagoras recognized certain musical consonances among the sounds produced by the hammers and *then* determined the physical cause. It is not the case that he chose certain mathematically satisfying relationships and then decided upon that basis to label those sounds consonant (which is contrary to what certain modern commentators have asserted). This fact is significant because it demonstrates that the trajectory of Pythagorean thought is not from the mathematical simplicity of certain relationships to the physical phenomena that conform to such simplicity, but rather from the physical phenomena to the simplest mathematical explanations that can account for them.

According to the standard version of the story, Pythagoras discovered that the musical interval known as the octave resulted from the ratio of 2:1, what the Greeks called the duple ratio. That is to say, when the first hammer was struck simultaneously with the second (the first hammer weighing twice as much as the second), the resultant sound was an octave. The octave is a very special sound. If you are sitting at a keyboard instrument (such as the piano) and you start from middle C, playing only white keys, then you will progress from C through D, E, F, G, A, and B before you return to C. That C is simultaneously different from and similar to the original C. We hear those two different pitches as somehow being the same *note* (precisely what that note is depends on the musical context, as we saw above) but in different registers—the second sounds higher than the first. Ptolemy later noted this quality of sameness by referring to pitches an octave apart as *homophones* (*homo* meaning "same"). The special nature of the octave remains a vital concern of music theory and will play an important role in the narrative that follows.

Pythagoras realized that the interval of a fifth arose from the two hammers whose weights were in the ratio of 3:2, which the Greeks called hemiolic.[3] This is the interval, for instance, between C and the G above it. Pythagoras found the interval of a fourth between hammers whose weights were in the ratio of 4:3, which the Greeks called epitritic. This is the interval that can be found between C and the F above it or between G and the C above it. If we now imagine the full array of hammers, with the heaviest hammer creating the lowest-sounding pitch, we might arrive at the following:

> hammer 1 = 12 units; hammer 2 = 6 units;
> hammer 3 = 4 units; hammer 4 = 3 units

Thus the set of ratios would be 12:6:4:3. If we designate the lowest-sounding note as middle C (c1), then hammer 2 would produce an octave above that (c2), hammer 3 would sound out the G above the latter C (g2), and hammer 4 would produce the C above the G (c3). To clarify:

---

3. The term *hemiolic* relates to our musical term *hemiola*, which generally means three beats against two.

12:6 reduces to 2:1 => the octave c1 to c2
6:4 reduces to 3:2 => the fifth c2 to g2
4:3 is in lowest terms => the fourth g2 to c3

Furthermore:

12:4 (hammer 1 and hammer 3) reduces to 3:1 =>
the octave plus fifth c1 to g2
12:3 (hammer 1 and hammer 4) reduces to 4:1 =>
the double octave c1 to c3

Here we have all of the intervals that a Pythagorean would have considered consonant (i.e., all of the possible combinations of notes that would have been considered pleasant and relatively stable): the fourth (4:3), fifth (3:2), octave (2:1), octave plus fifth (3:1), and double octave (4:1).

Though we have arbitrarily chosen to move from the low to the high sounds, we could have just as easily gone in the other direction. That is, starting with the lightest hammer, now represented by 1 unit of measure, we might get c3. The next hammer would give us c2 an octave below. The third hammer would produce f1 a fifth below the second hammer. And the final hammer would give us c1 a fourth below the previous hammer. Now the proportions in total are: 1:2:3:4, or the exact reverse of 12:6:4:3 inasmuch as we have the same ratios with the terms reversed. (As we will see in the next section, neither of these proportions is what is represented by most explicit accounts of the Pythagorean discovery. However, since all later accounts—that is, those that include specific numbers—are designed to address the concerns of their eras, we have no real reason to prefer their choice of arrangement over our own.)

We can now make some observations. Under this system, there are only two types of ratio that allow for a consonant interval: the multiple ratio expressed as $xn:n$ (2:1; 3:1; 4:1) and the epimoric or superparticular ratio expressed as $n + 1:n$ (3:2; 4:3). Multiple and superparticular ratios were preferred for their relative simplicity; the Pythagoreans believed that beauty should be readily perceptible and simple relationships are more readily perceptible than complex ones.

But clearly it is not enough to say that consonant intervals arise from multiple and superparticular ratios or else 7:6 should produce a consonant interval, which it does not. Therefore, a further condition must be operating here.

Notice that the numbers involved in these ratios (when placed in the lowest terms) are 1, 2, 3, and 4. The sum of these numbers is 10. The Pythagoreans considered the number 10 a particularly special number. In part, 10 is special because it is here that things kind of start over again. Moreover, 10 is special because of the way Pythagoreans understood numbers. The Pythagoreans represented numbers by arranging pebbles in particular formations and indeed those formations were part of the meaning of the number. The number 10 was arranged in the following pattern known as the tetraktys of the decad:

O<br>
O O<br>
O O O<br>
O O O O

Thus the tetraktys of the decad creates (or better, simply is) an equilateral triangle (four units per side), which Pythagoreans call a triangular number. More to the point, when taken from top to bottom, the tetraktys demonstrates the move, central to Pythagorean thinking, from the unit to three-dimensional space: 1 represents a point (location without dimension); 2 represents the two points required to draw a line (length without width or depth); 3 enables one to draw a triangular plane (length and width without depth); and finally 4 represents the first solid that inhabits three-dimensional space. Hence, for the Pythagoreans, the harmony of the consonances maps the coming into being of the physical world.

As we have already witnessed, and shall see in greater detail below, the Pythagoreans thought of a number *not* as a measure or representation of an actual thing but rather as itself an actual thing. Sounds in some sense actually *were* numbers, not simply things that could be represented by numbers. For many of the

Pythagoreans, sounds had a kind of magnitude and it was this magnitude that revealed their true nature. This provides us with a provisional answer to our original question concerning whether we were to measure simply the intervals or the pitches in themselves. For a Pythagorean, the pitches themselves had or were numbers.

Now, as evocative as the story of Pythagoras visiting the blacksmith shop is, it is in actuality based on bad science. The consonances produced by items of differing weights (say, for instance, weights pulling taut strings) would *not* be in the ratios described above. The blacksmith story is false with respect to the physics of sound. However, in ancient Greece, it is far more likely that Pythagoras would have been working not with hammers but rather with string lengths in his demonstration, and string lengths will give rise to the ratios described above. In other words, if you have a taut string that produces the pitch c1 and you place a bridge beneath it, dividing it in half, and pluck it, you will hear the pitch c2 an octave above the open string. This is easily demonstrated on any guitar. Pythagoras himself probably used (and may have invented) what we call a monochord—a single string stretched over a wooden board on which measurements can be taken.

The inconvenient truth behind the physics of the blacksmith story, however, does not mean that the story as it was generally told ought to be disregarded. There is an important rhetorical design to the story as it exists. The Pythagoras of the legend is *not* experimenting with bits of string. He is simply walking about in the world and he *recognizes* order within the seemingly cacophonous atmosphere of the smithy. Indeed, in the version of the story recorded by Boethius in his *De institutione musica*, there was a fifth hammer present that did not create a consonant interval with any of the other four. When Pythagoras realized which was the fractious hammer, he cast it aside and then proceeded with his investigation. This is, to say the least, a most telling detail. From the beginning of serious musical consideration, dissonance was thought to be in the way of true musical understanding. Dissonance had to be controlled, managed, or simply cast out. Music for Pythagoras, and really for nearly every Western musical thinker until the early twentieth century, was based on the stability and order of consonance. All seeming disorder, all dissonance, had to be subsumed beneath consonance or better yet, as certain Renaissance theorists would later say, *reduced to* (from the Latin *reducere*, meaning "to lead back to") consonance. Thus we can see how easily for Pythagoras and his intellectual descendants (and nearly all of us are his intellectual descendants) the *physics* of sound can become the *metaphysics* of music.

## The Pythagorean Discovery and the Ancient Greek Musical System

For the more attentive reader, there should be a few questions lingering. One question concerns the exact numbers used in explicit accounts of the Pythagorean story if not 1:2:3:4 or 12:6:4:3. After all, these numbers have certain undeniable benefits. They encompass all of the possible Pythagorean consonances (including the octave plus fifth and the double octave) and they do not introduce any intervals that are *not* considered consonant by the Pythagoreans. If not these numbers, then which numbers did later Pythagoreans prefer and why? The second question involves the names of the various intervals. Why, for instance, would we say that the consonance produced by the ratio 3:2 is a fifth, or as the Greeks would call it the *diapente* ("through the five")? Through five of what? Why does the ratio 4:3 produce a fourth, or *diatessaron*, and why does the ratio 2:1 produce an octave, a term meaning "eight," or what the Greeks called a *diapason* ("through all")?

The answers to these questions may be discovered through recourse to the ancient Greek musical system. This ancient system is rather far removed from our own in significant ways, and therefore even a fairly thorough background in current music theory will be of little help here. The basis of the Greek system is the tetrachord, that is, a group of four strings. The outer strings are fixed (meaning they are always in the same relationship to each other) but the inner two strings may vary in tuning according to the genus of the piece being performed.

We can now see where the term fourth or *diatessaron* comes from: "through the four" refers to the interval that spans the lowest to the highest of the four notes of the tetrachord. For a Pythagorean, this outer frame of the fourth would have been tuned to the proportion 4:3. Then the inner notes would vary their tuning in accordance with the genus. There were three genera: the diatonic, the chromatic, and the enharmonic. The diatonic tetrachord consisted of a semitone and two tones from the bottom to the top. Therefore, in terms of our system of notes, a diatonic tetrachord would consist of B C D E or E F G A, if we restricted ourselves to the white notes of the piano. This is because the interval between B and C and the interval between E and F are semitones while the intervals between all of the other adjacent notes are whole tones. (This is, of course, why when you look at a piano, you will find no black keys between B and C or between E and F—there are no black keys in those positions because only a semitone separates those notes.)

The chromatic genus in ancient Greece differs from our concept of the chromatic. In this genus, the tetrachord still had only four notes but they would be tuned so that the first two intervals were semitones and the last would be the approximate equivalent of what we would call a minor third. Thus a chromatic tetrachord would consist of what we might think of as B C C# and E. Notice that the framing relationship of B to E would remain unaltered. Finally, the enharmonic genus, once again, contains four notes, but now the arrangement is two quarter-tones and an approximate equivalent to our major third. Our common ideas concerning musical pitch do not allow for an easy representation of this genus, but it would consist of what we might think of as B, a quarter-tone above B, a quarter-tone above that (basically equivalent to our C), and the E. Once again, the frame of a fourth (ratio 4:3) would remain the same.

The ancient Greeks seem to have imagined their tuning system as though it were a many-stringed lyre. Certain notes of the lyre remain fixed (like the framing fourth of our tetrachord) while others are changeable according to genus. The Greeks constructed the systems themselves out of a joining of tetrachords. Tetrachords could be assembled in a conjunct manner (i.e., with the highest string of one tetrachord being identical with the lowest string of the next) or in a disjunct manner (where the tetrachords share no strings and are separated by the distance of a tone). For instance, one could combine our tetrachord (in the diatonic genus for the sake of simplicity) with another in a conjunct manner as follows:

**B** C D **E** F G A

where the large, bold-faced typeset represents the fixed notes.

We can visualize this more easily as follows:

first tetrachord

**B** C D **E**

**E** F G **A**

second tetrachord

Notice the overlap between the two tetrachords. They share the string tuned to E.

A disjunct connection would appear as follows:

first tetrachord second tetrachord

**E** F G **A** **B** C D **E**

Here there are no shared strings. The tone that separates the highest note of the first hexachord (A) from the lowest of the next hexachord (B) was termed the "tone of disjunction."

How was the tone of disjunction tuned? And, to return to our earlier question, what were the actual numbers used in later accounts (such as the account provided by Boethius) of the Pythagorean discovery? To answer that, notice that by combining tetrachords, we have expanded the kinds of intervals available to us beyond the framing limits of the fourth. Indeed, if you look at the highlighted notes in our example of a disjunct combination of tetrachords, you will notice that from E (the lowest note of the first tetrachord) to B (the lowest of the second tetrachord) or from A (the highest note of the first hexachord) to E (the highest note of the second tetrachord) there lies the interval of a fifth, or a diapente (i.e., "through five [strings]"). Indeed in the disjunct combination of two tetrachords of the same genus (here diatonic) each note in the second tetrachord is a fifth above its corresponding note in the first tetrachord.

Therefore, the whole tone was defined on the basis of the tone of disjunction. That is to say, the whole tone was defined as the remainder after the subtraction of the fourth (4:3) from the fifth (3:2). Since in subtracting ratios, one divides and in dividing, one cross-multiplies, we arrive at the following:

$$\text{fifth} - \text{fourth} = 3{:}2 \,/\, 4{:}3 = 3 \times 3{:}4 \times 2 = 9{:}8$$

Thus the Pythagorean ratio for the whole tone is 9:8, what the Greeks called the epogdoic ratio. (Notice that 9:8 is a superparticular or epimoric ratio but it does *not* produce a consonance; a Pythagorean would explain this fact by noting that the terms of its ratio fall outside the numbers contained in the tetraktys of the decad.) This, now, allows us to see the measurements for the diatonic tetrachord quite clearly. The frame is a fourth 4:3 and each whole tone is 9:8. Therefore, the semitone (which is not simply half the distance of a whole tone since it is mathematically impossible to evenly divide the ratio 9:8) is what is left over when you subtract two whole tones from the fourth. Hence:

fourth – two whole tones = 4:3 / 81:64
(the sum of two whole tones) = 256:243

Thus the Pythagorean semitone is 256:243. This interval was called the diesis by Pythagoras's follower Philolaus (among others) and was referred to as the leimma by Plato and other commentators. Thus the diatonic tetrachord (from low to high) consists of a diesis followed by two epogdoics (or whole tones) within the framework of an epitritic ratio (or the fourth—4:3).

Now we are ready for the payoff inasmuch as we can now understand the actual numbers used in most accounts of the Pythagorean discovery story. At least since Philolaus (and perhaps even before his time), the Pythagorean discovery was presented in such a manner that it provided a framework of fixed notes within the Greek tuning system. Thus the hammers are said to have been in the following proportion (with the hammer containing the most units of weight producing the lowest-sounding note):

hammer 1 (12 units) : hammer 2 (9 units) :
hammer 3 (8 units) : hammer 4 (6 units)—
or, 12:9:8:6

In this presentation, the notes are framed by an octave, or diapason (12:6 reducing to 2:1) between hammers 1 and 4. Between hammers 1 and 2 and between hammers 3 and 4 lie intervals of a fourth, or diatessaron (12:9 and 8:6 both reducing to 4:3). Between hammers 1 and 3 and between hammers 2 and 4 lie intervals of a fifth, or diapente (12:8 and 9:6 both reducing to 3:2). Finally, the epogdoic ratio of 9:8 lies between the sounds produced by hammers 2 and 3. Thus, if we take hammer 1 to produce the pitch c1, we get the following array: **c1 f1 g1 c2**. This seems to be the array desired by the earliest authorities describing the Pythagorean discovery (that goes back to, perhaps, Sextus Empiricus).

But why this configuration? Our initial array had so many wonderful advantages. Not only did it include nothing but consonances but it also included *all* of the Pythagorean consonances! This other proportion (12:9:8:6) fails to include the double octave (4:1) and the octave plus fifth (3:1) *and* it includes an actual dissonance (9:8)! However, what this particular proportion accomplishes is a clear delineation of the layout of the tuning system.

Take a careful look at figure 6-1. This represents what is called the Greater Perfect System in ancient Greek music theory (we need not concern ourselves with the Lesser Perfect System, for our purposes). The first column lists the four tetrachords of this system (the dotted lines surround strings that are included in both adjacent tetrachords). The next column lists the Greek names of the strings (imagine the system as a large lyre) and the third column denotes the approximate relationships between the ancient Greek strings and our note names.[4] Do not be put off by the foreign words; they simply name the strings according to their position in the tetrachords. (They are also kind of cool.)

Starting from the top, we find tetrachord hyperbolaion consisting of the following four strings: nete hyperbolaion, paranete hyperbolaion, trite hyperbolaion, and nete diezeugmenon. Thus from the lowest to the highest note of this tetrachord, we find a fourth (4:3). The next tetrachord, tetrachord diezeugmenon, consists of the following strings: nete diezeugmenon, paranete diezeugmenon, trite diezeugmenon, and paramese. Notice that tetrachord hyperbolaion and tetrachord diezeugmenon share a string (namely, nete diezeugmenon) and are therefore conjunct tetrachords.

4. Note that these note names are only approximately accurate and only apply to the diatonic genus.

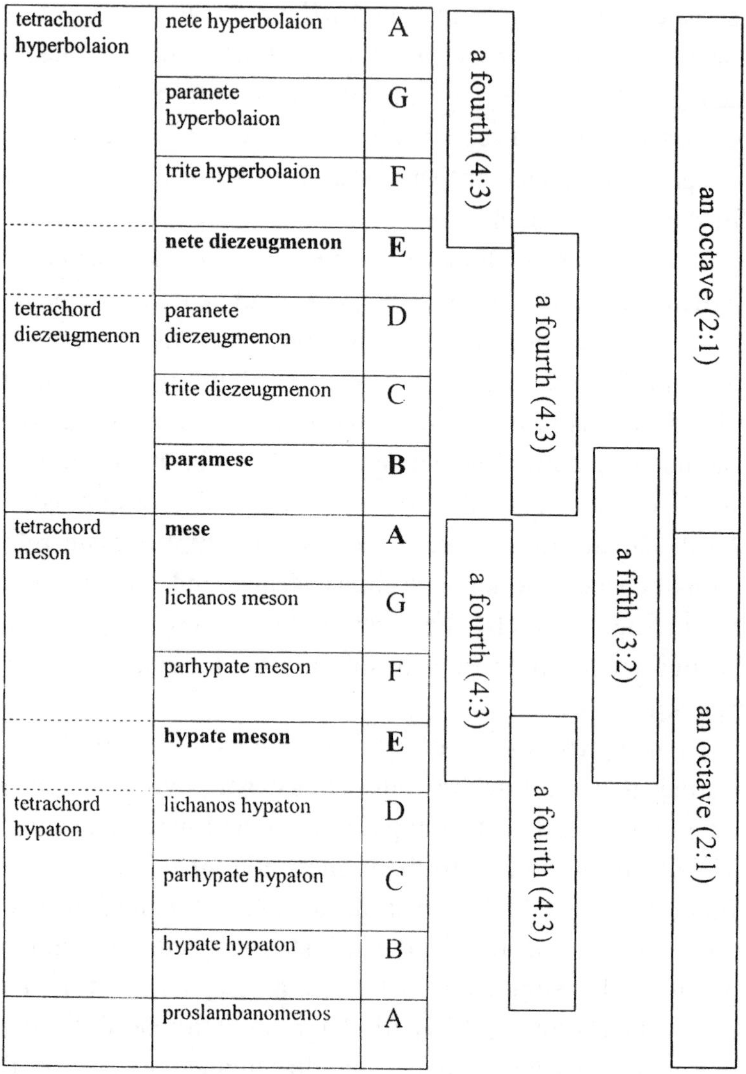

|  |  |  |
|---|---|---|
| tetrachord hyperbolaion | nete hyperbolaion | A |
| | paranete hyperbolaion | G |
| | trite hyperbolaion | F |
| | **nete diezeugmenon** | **E** |
| tetrachord diezeugmenon | paranete diezeugmenon | D |
| | trite diezeugmenon | C |
| | **paramese** | **B** |
| tetrachord meson | **mese** | **A** |
| | lichanos meson | G |
| | parhypate meson | F |
| | **hypate meson** | **E** |
| tetrachord hypaton | lichanos hypaton | D |
| | parhypate hypaton | C |
| | hypate hypaton | B |
| | proslambanomenos | A |

a fourth (4:3)

a fourth (4:3)

a fourth (4:3)

a fourth (4:3)

a fifth (3:2)

an octave (2:1)

an octave (2:1)

**Figure 6-1**
**The Greater Perfect System.**

The next tetrachord is tetrachord meson and it contains the following strings: mese, lichanos meson, parhypate meson, and hypate meson. Notice that tetrachord meson and tetrachord diezeugmenon do not share any strings in common and are therefore disjunct (indeed, the word "diezeugmenon" means disjunct). Finally, tetrachord hypaton contains the following strings: hypate meson, lichanos hypaton, parhypate hypaton, and hypate parhypate hypaton, and hypate hypaton. Thus tetrachord hypaton and tetrachord meson share the string hypate meson and are therefore conjunct tetrachords. At the bottom of the system is a string that belongs to no tetrachord, proslambanomenos, added so that the entire system spans two octaves (the largest Pythagorean consonance).

Now look carefully at the central portion of the Greater Perfect System, specifically at the two centrally located tetrachords (meson and diezeugmenon), paying attention only to the fixed notes (in bold). From the lowest note of tetrachord meson to the highest note of tetrachord diezeugmenon is the interval of an octave. Each tetrachord individually spans a fourth. From the lowest string of tetrachord meson to the lowest of tetrachord diezeugmenon is a fifth as is the distance between the highest notes of each tetrachord. Finally, the distance separating the highest note of tetrachord meson (the A, or mese) and the lowest note of tetrachord diezeugmenon (the B, or paramese) is a whole tone (or the epogdoic ratio of 9:8).[5] This arises, of course, because the *strings* are what are really being counted.

Don't worry if you find you have gotten a bit lost in the alphabet soup of the ancient Greek names for the strings. The important element here is the arrangement of intervals in the central portion of the system. Even if you did not follow all of the details in that account, we can now see why the Pythagoreans set out the proportion as 12:9:8:6. It accounts for the ratios of the three main consonances (the octave, fifth, and fourth) *and* it presents them in the arrangement of the fixed notes of the central portion of the Greater Perfect System (thereby clearly designating the interval of the tone

5. Notice, for those readers not as familiar with music, that this means that a fifth plus a fourth equals an octave (as though $5 + 4 = 8$)!

of disjunction as 9:8). Thus what these authors claim that Pythagoras "discovered" is not simply the ratios of the consonances but also the proper ordering of the Greek tonal system—not bad for a walk downtown! As we shall now see, however, this was hardly the primary value of the discovery for Pythagoras and his followers. Indeed, Pythagoreans, strictly speaking, were not all that interested in music as a human practice. Their interest was strictly in what observations concerning the consonances and musical harmony could tell us about the much larger philosophical notion of *harmonia*—that pervasive quality of the cosmos that holds things together, allows us to understand the world and things around us, and maintains the relationship between our bodies and our souls.

## *Discordia concors*: Music and the Pythagorean Cosmos

Pythagoras's musical discovery was confirmation of, or perhaps even the launching pad for, a deeper Pythagorean insight: that the universe and everything in it could be explained through numbers. After all, what was it that made those consonances so appealing if not the numerical proportions underlying them? Alter the proportions and concord becomes discord. Moreover, a consonance is a very special type of phenomenon for the Pythagoreans. Nichomachus of Gerasa, an ardent Pythagorean, made the intriguing and highly influential claim that a consonance was the fusion of two sounds into a single entity whereas the constituent notes of a dissonance remained in their integrity. Thus dissonance is a type of metaphysical nullity, a pure nothing (reinforcing the observation above that dissonance requires careful handling if not outright dismissal). Consonance, however, takes different things and makes them one, or, as a later generation of music theorists would put it—*discordia concors* (concord made out of diversity).

Now what can it be that holds the planets in their orbits, that organizes matter, that maintains the alternating seasons, and that controls the interaction of souls and bodies, if not a similar phenomenon to what gives rise to musical consonance? This is what the Pythagoreans ultimately meant by the term *harmonia*—the underlying numerical order that holds all things together.

For the Pythagoreans, the ultimate background notions of existence were the unlimited and the limit. The limit is that which orders chaos. It is because of the ordering capacity of the limit that we are able to exist, to have perceptions that are reliable, and to have knowledge that is accurate. That knowledge is of the system of order itself for that is Truth. Thus Pythagoras's insights into the mathematical order of music seem to have led him to perceive mathematical order everywhere.

It is no surprise therefore that the Pythagoreans were devotees of the god Apollo (indeed, some of Pythagoras's followers believed him to be an incarnation of the Hyperborean Apollo). Apollo is the god of order and reason. His temple displayed inscriptions of sayings such as "Observe limit" and "Nothing too much." He was the god of what the Greeks called *logos* or what in Latin was termed *ratio*. These are notoriously difficult terms to translate but they cover a range of meanings, including "the intelligible, the measurable, the rational, and the determinate." *Logos* is concerned with the proportions among things and between parts and wholes.

The universe, for the Pythagorean, was rational—its rationality being underwritten by number. Thus the Pythagorean Philolaus declares in one of the famous fragments of his writings that all things that are known and can be known have number. Indeed he goes on to insist that things not possessing number are simply unknowable. Pythagoras himself may have been the first person to use the term *kosmos* to register the idea of an ordered universe. *Cosmos*, another difficult term to translate in its Greek sense, implies a combination of perfection, proportion, and beauty that arises from such order. Just as the musical consonances arose from a mathematical order, so did the living cosmos. Indeed, some Pythagoreans went so far as to propose a hypothetical tenth heavenly body, called the Counter-Earth, so that the number of heavenly bodies would come to the tetraktys of the decad: the celestial sphere (the outermost portion of the cosmos to which the stars were affixed), the sun, the moon, Mercury, Venus, Mars, Jupiter, Saturn, Earth, and Counter-Earth.

The relationship among these heavenly bodies depended upon the same kinds of proportions as those found among the consonances or within the Greater Perfect System. This is the concept often referred to as "the Harmony of the Spheres." Some Pythagoreans believed that this cosmic harmony was actually audible; the reason we do not notice it is either because it is too soft (an idea dismissed by other Pythagoreans based on the massive size of the heavenly bodies) or because we are just too used to it. Indeed, members of the Pythagorean cult held that once you had achieved a higher understanding of the cosmos, you would finally hear the cosmic music.

Becoming aware of the cosmic harmony meant that you were also becoming aware of the harmony between your body and your soul. As Plato attempts to demonstrate in his difficult but rewarding treatise *Timaeus*, the same basic proportions that hold the cosmos together are also the proportions that comprise the makeup of our souls. This is because, for Plato as for the more orthodox Pythagoreans, the cosmos simply is a soul—what is referred to as the World Soul—and our individual souls are but a portion of that larger soul. This then reveals the reason for the Pythagoreans' concern with these seemingly abstruse matters. Like many ancient Greeks, the Pythagoreans believed the dictum "Like is known by like." They did not, however, interpret that maxim the way many other Greek thinkers did. For instance, Pindar gives a pretty standard interpretation of this saying when he declares that mortal men should limit themselves to mortal thoughts and should not be concerned with notions of immortality, ultimate truth, and the organization of the universe. The Pythagoreans, however, believed otherwise.

According to the Pythagoreans, a human being is a mixture of matter (the body) and the divine (the soul). Whereas the body is mortal, the soul is not. It participates in the eternity of the World Soul. The World Soul simply *is* the body of knowledge pertaining to its organization (i.e., it is the cosmos). The Pythagoreans, therefore, did not take the maxim "Like is known by like" to imply a sense of limit or stasis. Rather they held a dynamic notion of the saying. Instead of being limited to the mortal side of our nature, our participation in the divine allows us to transcend it. Thus by coming to know the order behind the universe, the soul, the virtues, and the music, we *augment* the divine part of our nature. We know more about who we are. Pythagoras, who some claimed to have coined the term "philosophy," believed that our road to salvation was laid through understanding and our capacity for knowledge makes possible our assimilation with the divine.

Music participated directly in the betterment of humankind not simply by providing the cartography of the cosmos but also by helping to "retune" those souls that fell into a dissolute state. Thus Pythagoras was said to have cured drunkenness, leprosy, and excessive anger simply by having music played employing specific intervals. Music therapy is thus a far older practice than many of us realize.

It should come as little surprise, therefore, that music, arithmetic, geometry, and astronomy were all thought of as being closely related. The philosopher Boethius (ca. 480–ca. 525) apparently coined the term *quadrivium* (literally, the "four ways") to account for these four mathematical sciences. Arithmetic deals with number in itself—that is, with quantity or multitude. Geometry deals with planes and solids or magnitudes (volumes). Music treats numbers in relationships (ratios) and astronomy deals with magnitudes in motion (the revolving constellations, for instance). Boethius divided the study of music into three parts: *musica mundana* (the study of the harmony of the spheres, the changes of the season, etc.); *musica humana* (the study of the relationships among the soul and the body); and sounding music. Only the latter pertains to what we would recognize as music today. Indeed aspects of the former two have little or nothing to do with sound at all—demonstrating that for Boethius the defining aspect of music was not necessarily sound but the relationships of measurable things.

Boethius also divided the types of people involved with music into three kinds: the performer (basically a slave who does what others demand and is overly tied to the body); the composer (who has an intuitive understanding of music but lacks actual knowledge); and the true musician (one who understands the relationships inherent in the study of music and knows how those relationships map onto the soul and the cosmos). Thus for Boethius, the real musician does not perform and does not compose (necessarily) but rather the real musician is the one who *knows*. Tell that to your musician friends!

## Pythagoras and Later Musical Thinking

Pythagorean thought regarding music did not simply subside after antiquity. Indeed, Pythagorean beliefs experienced a notable revival during the Renaissance and the early modern period—particularly in the writings of such figures as Marsilio Ficino, Marin Mersenne, and Robert Fludd; all of these writers espoused some version of the harmony of the spheres. However, music theorists intrigued by Pythagorean claims were forced to alter certain notions by the exigencies of a later musical practice. Most important, musicians no longer limited the consonances to the octave, fifth, fourth, octave plus fifth, and double octave. These intervals (along with the somewhat contentious octave plus fourth) were now recognized as *perfect* consonances (a reminder of their ancient origins and the relative simplicity of their ratios), while composers enriched the musical textures of their compositions through the use of *im-*

*perfect* consonances: namely, the thirds and sixths (both major and minor).[6]

The addition of the imperfect consonances raised both practical and theoretical difficulties. Pragmatically, the major third produced through strictly Pythagorean measurements would be of the ratio 81:64 (arrived at through the sum of two Pythagorean whole tones—9:8 + 9:8 = 81:64). This is a rather harsh interval; the notes seem to have struck the Renaissance ear as being just slightly too far apart. However, if one were to alter the major third to make it more pleasing, then that would give rise to different sizes of whole tones. This is precisely the solution to the difficulty that one finds in the writings of Fogliano and more famously in Gioseffo Zarlino's widely read *Le istitutioni harmoniche*. Zarlino proposed the following tuning system, which he claims was derived from one of the tuning options proposed by Ptolemy (thus retaining its ancient credentials), called the "syntonic diatonic."

| C | | D | | E | | F | | G | | A | | B | | C |
|---|---|---|---|---|---|---|---|---|---|---|---|---|---|---|
| | 9:8 | | 10:9 | | 16:15 | | 9:8 | | 10:9 | | 9:8 | | 16:15 | |

This tuning system has the following advantages: the C to F and the G to C are perfect fourths (4:3); the C to G and F to C are perfect fifths (3:2); all three major thirds (C to E, F to A, and G to B) are of the pleasing ratio 5:4; two of the three minor thirds (A to C and B to D) are also of the pleasing ratio 6:5. The only real drawbacks of the system are the minor third between D and F holding the ratio 32:27

6. The problem with the octave plus fourth is that its ratio is 8:3, which is neither a multiple nor a superparticular ratio. Therefore, even though generally the octave was thought *not* to alter the quality of an interval to which it was added, in this case adding an octave and a fourth created a *dissonant* ratio numerically that sounded *as though* it were a consonant ratio. Such a realization did not deter the early Pythagoreans from their understanding of consonance but rather confirmed their belief that reason ought not to subject itself to the deceitful nature of the senses!

(a comma—81:80—larger than 6:5), and the fifth between D and A of the ratio 40:27 (a comma larger than 3:2).

Notice that the optimal intervals within Zarlino's tuning system are as follows:

octave: 2:1
fifth: 3:2
fourth 4:3
major third: 5:4
minor third: 6:5

So far, these are all superparticular ratios (except for the multiple 2:1) just like in the Pythagorean notion of consonance. However, the ratios are not limited to the first four numbers for their constituent parts. Rather the ratios are made up of the first *six* numbers. Moreover, Zarlino proposed a rather Pythagorean argument to justify the departure from the tetraktys of the decad. The number six, Zarlino notes, is the first *perfect* number inasmuch as a perfect number is one for which the factors of the number (excepting the number itself) add up to the number, as follows: the factors of 6 are 1, 2, and 3; and 1 + 2 + 3 = 6.

Since any Pythagorean would have to agree that music arises from the perfection of the numbers, Zarlino's claim that the consonances are bounded by the parts of the first *perfect* number would seem to outmaneuver the Pythagorean preference for four. Thus Zarlino replaced the tetraktys of the decad with what he termed the *senario* (the senary number) and made many of the same cosmic claims for its significance as the Pythagoreans had made for the cosmic significance of four.

Zarlino recognized that this move did not get him out of all of his difficulties. The major and minor sixths were also considered consonances, but in Zarlino's tuning system they were not quite as simple with respect to their ratios as the other consonances. The major sixth is of the ratio 5:3—both of the numbers are within the *senario* but it is not a superparticular ratio (nor is it a multiple ratio). More troubling is the minor sixth with the ratio 8:5. Not only is this ratio not superparticular, its components include the number eight, which falls outside of the *senario*. Zarlino addressed this dilemma by claiming that while *in actuality* the eight is outside of the *senario*, it is within the *senario* in potentiality (*in potenza*). To justify this seemingly absurd argument, he used as recourse the powerful philosophical ideas of Aristotle—but that is another story altogether.

At the end of the Renaissance, Vicenzo Galilei, the father of the famous mathematician and inventor Galileo, proved through an experiment that the Pythagorean story of discovery in its most typical version was simply false. Most authorities assert that after observing the hammers in the smithy, Pythagoras experimented at home with weights in the same proportions as the hammers suspended from strings. Thus, supposedly, he was able to re-create the same consonances. This experiment was supposed to demonstrate that consonances simply *are* the result of certain numerical relationships. Galilei tried it out for himself and discovered that the production of a perfect fifth (3:2 when taking string lengths into account) required weights in the ratio of 9:4; an octave required weights in the ratio of 4:1; and a fourth required 16:9. In other words, when producing intervals from weights suspended on strings, the parts of the ratios had to be squared. When producing intervals from volumes of concave bodies (like organ pipes), the parts of the ratios had to be cubed.

For Galilei this meant that there was no consistent and necessary relationship between the ratios produced as a measure of the sound-producing objects and the actual intervals that arose from the sounding bodies. Thus all of the numerical theorizing from antiquity to Zarlino, for Galilei, was just so much hot air.

Numbers, nonetheless, would not be banished from music theory for very long (if indeed we can claim that they ever were banished). In fact, Galilei's own son was integral in establishing the fact that pitch arose on account of the frequency of the sound and that such frequency could be measured. But even more significant was Joseph Sauveur's discovery, published in 1701, of the overtone series. This is the phenomenon that causes a sounding body to not only vibrate at a given frequency (referred to as the fundamental) but also at twice that frequency, three times that frequency, and so on. This gives rise to the following intervallic pattern for a string tuned to the pitch A at the frequency of 110 cycles per second:

A (110), A (220), E (330), A (440), C# (550), E (660)

Most writers agree that human beings generally cannot hear beyond the fifth overtone (the sixth partial). This insight into the physics of sound reveals that the octave may be considered so special because we are hearing it on some level *every time* we hear a clear, pitched tone. The fifth (another very important interval) also appears within the overtone series (the E above the A).[7] Finally, notice that the first six partials produce a major triad (A C# E). The major triad was already noted by Zarlino to be of extreme significance; he called it the "perfect harmony."

However, as music moved away from the older modal systems of organization toward what we now call the *tonal* system of organization, the major triad would move from being a very special sonority to being the foundation of the entire system. Thus the most important early theorist of tonality, Jean-Phillipe Rameau, would refer to the major triad as the "Chord of Nature." Of course, by this point in history (the early eighteenth century), we are rather far removed from Pythagoras's original conceptions of music and consonance. But it is nevertheless revealing that the foundations of the modern tonal system maintained a debt to the Pythagorean insight that discerned some hidden order behind the phenomenon of the consonance. In a very real sense, when it comes to music, we are all still Pythagoreans.

---

7. Notice that the fourth does not appear directly above the fundamental. Perhaps it is understandable, consequently, that the fourth has always been a troubling consonance for theorists and practitioners. For instance, in two-part counterpoint, the interval of a fourth is treated as a *dissonance* that must be resolved.

# Chapter 7

# The Pythagorean Theorem in Fractal Art[1]

During World War II, a Dutch engineer hired by the Germans quietly took time to pursue his interest in geometric drawings. On the same drawing board on which he was to design submarines, Albert E. Bosman (1891–1961) pursued a much more abstract configuration. He constructed squares on the sides of right isosceles triangles and was led by curiosity to investigate what kind of figure would be created when the pattern was extended, placing new triangles on the squares, then squares on the triangles, and so on. He expanded the pattern until the details became too intricate and small on his 60 by 85 cm (roughly 23.6 in. by 33.5 in.) sheet of drawing paper.

Bosman was unaware that he had forged the foundation for a new class of fractals now termed "Pythagorean trees." Fractal geometry was then in its infancy, and it was to receive its current popularity only a few decades later with the advent of computers. Since fractals are endlessly repeating figures, the work that took Bosman days to accomplish by hand drawing can be done with much more accuracy in a few minutes with the aid of computers.

1. This chapter was written by Dr. Ana Lúcia B. Dias and Dr. Lisa DeMeyer, associate professors in the Department of Mathematics at Central Michigan University.

Bosman's drawing appeared in his 1957 book, *Het Wondere Onderzoekingsveld der Vlakke Meetkunde* (Geometry in the Plane: A Miraculous Field of Research), and then in a 1962 issue of the Dutch magazine *Pythagoras*. The drawing not only influenced other artists, such as Flemish-Belgian Jos de Mey (1928–2007),[2] but those interested in fractals as well.

## Pythagorean Trees

While different fractals have their idiosyncrasies, two concepts are common to all fractals: *self-similarity* and *recursion*. Fractals are obtained by applying a rule of choice to an initial figure, which we will call a "seed" object, and then recursively repeating the procedure (or rule) on the resulting figures or "outputs." The resulting figures will present some degree of self-similarity: whenever we zoom in (i.e., enlarge) on different portions of the resulting figure, we will see reduced copies of the whole figure.

Pythagorean trees are fractals obtained from the classical diagram—which we saw earlier—that illustrates the Pythagorean Theorem: a right triangle with squares constructed on its sides (figure 7-1). The familiar diagram is used as the fractal's "seed," or "Stage 0."

2. See Bruno Ernst's book *Bomen van Pythagoras: Variaties van Jos de Mey* (Trees of Pythagoras: Variations by Jos de Mey) (Amsterdam: Aramith, 1985).

**Figure 7-1**
**Stage 0 of a Pythagorean tree.**

The rule that we will use to generate the fractal is as follows: On each of the smaller squares (squares on the legs of the right triangle) we will construct two squares, forming right angles at their common vertex. They will be placed on the squares at the same angles formed by the sides of the original triangle, thus creating similar, reduced copies of the original diagram. There are two ways to do this. We can keep the original position of the angles or we can swap them (figure 7-2). Whichever the choice, the construction of the fractal continues by repeating the chosen procedure.

**Figure 7-2**
**Two possible ways of proceeding with the construction of Pythagorean trees: In the left-hand diagram, the angles of the triangles have their relative positions maintained. In the right-hand diagram, the angles have their position alternated, that is, the smaller triangles are reflected, reduced versions of the original triangle.**

If we choose to keep the original arrangement of angles, the result is a "leafy" Pythagorean tree. Quite a different result is obtained if we choose to alternate the position of the angles at every iteration. The tree resembles then a coniferous shrub. Figures 7-3 and 7-4 show Stage 2 of the construction of leafy and coniferous Pythagorean trees, respectively.

Depending on the measurements of the acute angles in the original configuration, the resulting trees can look quite different. Figures 7-5 and 7-6 show two different leafy trees, one having the squares set at angles measuring 90°, 53° and 37°, and the other at angles 90°, 62°, and 28°. The coniferous versions obtained with the same angle measurements are shown in figures 7-7 and 7-8.

**Figure 7-3**
**Stage 2 of the construction of a leafy Pythagorean tree.**

**Figure 7-4**
**Stage 2 of the construction of a coniferous Pythagorean tree.**

**Figure 7-5**
**Stage 7 of the construction of a leafy Pythagorean tree. The acute angles in the right triangles are 53° and 37°.**

**Figure 7-6**
**Stage 7 of the construction of a leafy Pythagorean tree. The acute angles in the right triangles are 62° and 28°.**

**Figure 7-7**
**Stage 7 of the construction of a coniferous Pythagorean tree. The acute angles in the right triangles are 53° and 37°.**

**Figure 7-8**
**Stage 7 of the construction of a coniferous Pythagorean tree. The acute angles in the right triangles are 62° and 28°.**

What is the area of a Pythagorean tree? Let's consider the area of the square on the hypotenuse of the triangle in Stage 0 to be our unit of area. The area of the fractal at that stage of the construction is thus 2—since by the Pythagorean Theorem the area of the two squares on the legs of the right triangle, when added, will equal that of the square on the hypotenuse (figure 7-1). The first iteration adds four squares to the construction (figure 7-2). By using the Pythagorean Theorem, we find their area adds up to 1 unit. Generally, the $n$th iteration in the construction originates $2^{n+1}$ squares with a total area of 1. Thus the area of the tree seems to grow without bounds. However, sooner or later and depending on the angles used in the construction, some of the squares start overlapping.

In particular, if the original triangle is an isosceles right triangle, there are no overlaps up to the third iteration (figure 7-9). From then on, the tree grows "inward" as well as "outward," but never past the boundaries of a well-determined rectangle, as Bosman himself had noticed. For example, if we start with an isosceles right triangle in which the hypotenuse is one unit long, we can see that the fractal will be, at any stage of its construction, confined within a 4 by 6 rectangle (figure 7-10). This shows that its area is no greater than 24 square units, and thus finite. This is one of the counterintuitive properties of fractals. Although they are obtained by an infinite process, they may turn out to have a finite area.

**Figure 7-9**
**The three initial iterations in the construction of a Pythagorean tree with isosceles right triangles. Notice no overlapping occurs up to this stage.**

**Figure 7-10**
**The grid shows the area of this Pythagorean tree is no greater than 24 square units (four grid squares equal one unit of area).**

## A New Fractal

In chapter 2, Demonstration 18, we presented a clever proof of the Pythagorean Theorem using the diagram in figure 7-11. Remember we started with congruent right triangles $ABC$ and $DGC$, where $AC = GC$, and placed them as shown in figure 7-11.

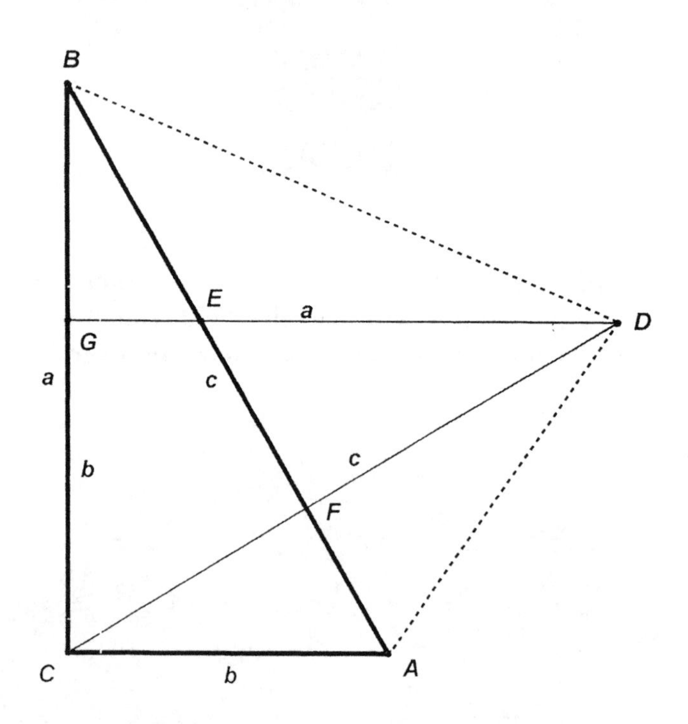


**Figure 7-11**

This diagram provides us with the inspiration for a new fractal construction. We will start our fractal with a nonisosceles right triangle. Our "generative" rule (i.e., our rule to generate a fractal) will consist of the same procedure that led to the diagram in figure 7-11, but followed by one more step. We are going to remove from the diagram the quadrilateral in the middle (*GEFC*), leaving just the three smaller right triangles in the diagram. Figure 7-12 shows the procedure.

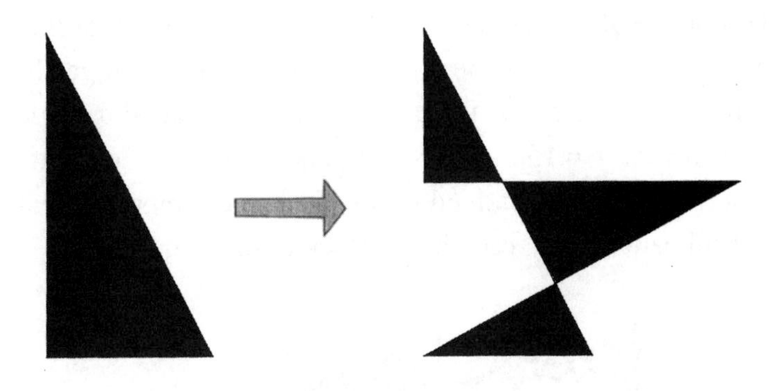

**Figure 7-12**
**Stages 0 and 1 of the construction of the fractal.**

We now apply the same procedure to all right triangles in the diagram. The next stage of the construction of the fractal is shown in figure 7-13.

**Figure 7-13**
**Stage 2 of the construction of the fractal.**

To obtain subsequent stages of this fractal, we repeat the generative procedure in all the right triangles at each stage. Figure 7-14 shows Stage 5 of the fractal, using as the original triangle that of sides $a = 8$, $b = 15$, and $c = 17$. As the number of iterations increases, the figures obtained become more and more intricate, in a beautiful pattern that resembles a flock of birds in flight.

**Figure 7-14**
**The fractal after 5 iterations. Original triangle is (8, 15, 17).**

Notice that in figure 7-14, in which we used the (8, 15, 17) triangle (and therefore the nonright angles have measures 61.93° and 28.07°), there is a small overlap of two triangles. The overlap is carried on to subsequent stages (see figure 7-15). This does not compromise self-similarity, though, as copies of the original diagram can still be seen in different portions of the fractal.

We may ask ourselves what happens if we change the angles of the original right triangle. Figures 7-15, 7-16, and 7-17 show the

Stage 5 configuration of three distinct versions of the fractal, with each using different original triangles. They show that the greater the difference between the nonright angles (that is, the further the triangle is from being isosceles), the more overlaps occur in the fractal.

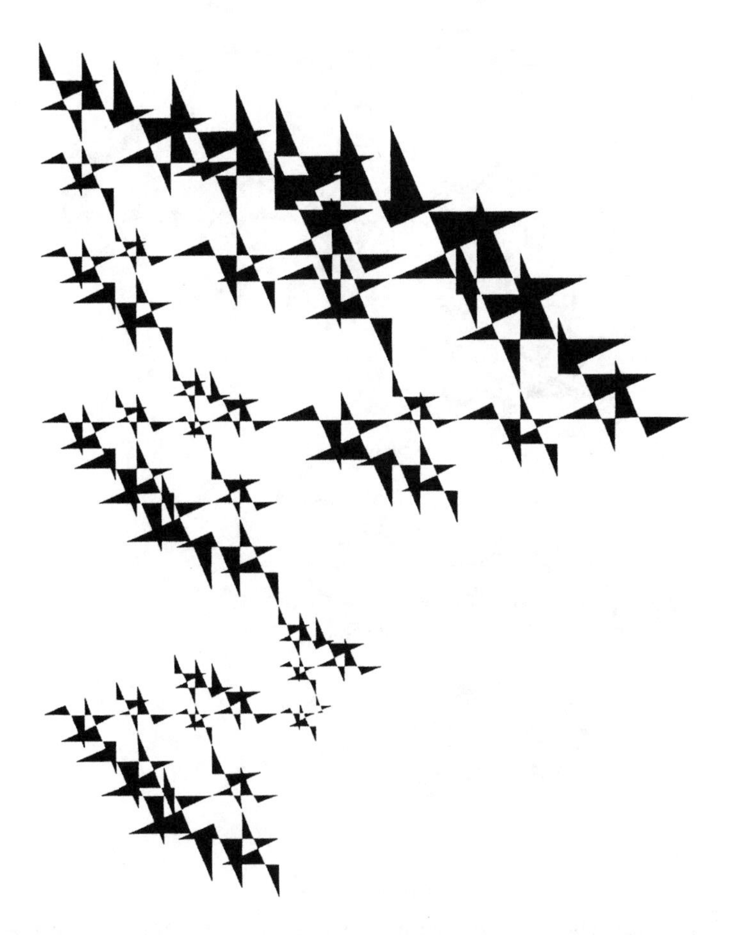

**Figure 7-15**
**Stage 5 of the construction of a fractal with nonright angles measuring 24° and 66°.**

**Figure 7-16**
**Stage 5 of the construction of a fractal with nonright angles measuring 30° and 60°.**

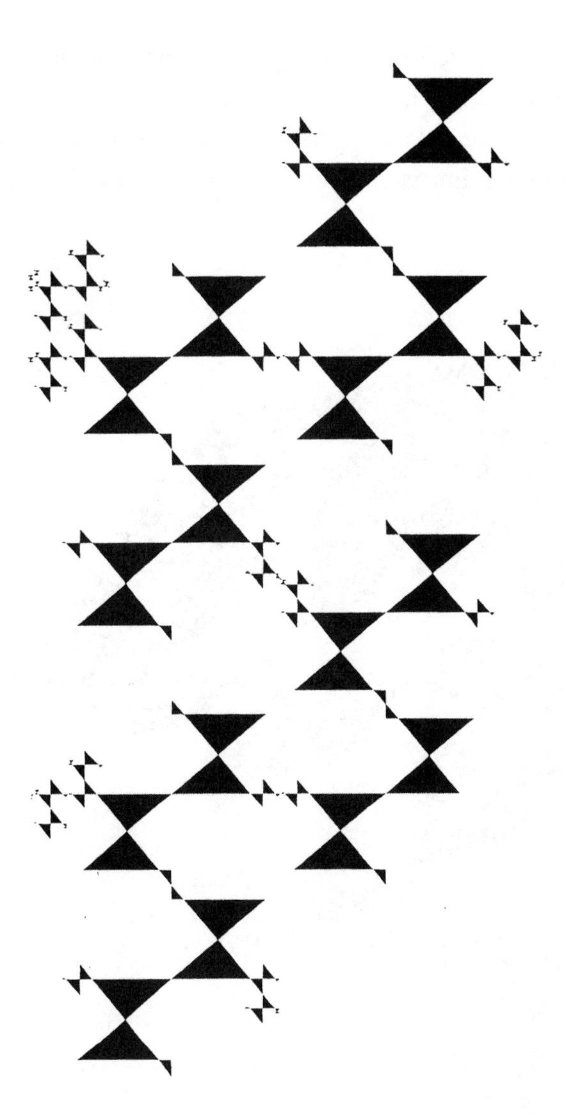

**Figure 7-17**
**Stage 5 of the construction of a fractal with nonright angles measuring 40° and 50°.**

If we continue the construction of those fractals, it becomes more evident that a slight change in the angles of the original triangle will produce huge changes in the overall appearance of the fractals. Figures 7-18, 7-19, and 7-20 show the Stage 8 configuration of the same fractals shown in the preceding three figures. Remember the only thing we did was to use different angles at Stage 0. But the effect of small changes accumulates as we perform more and more iterations, and soon the resulting figures don't even seem to be part of the same family of fractals.

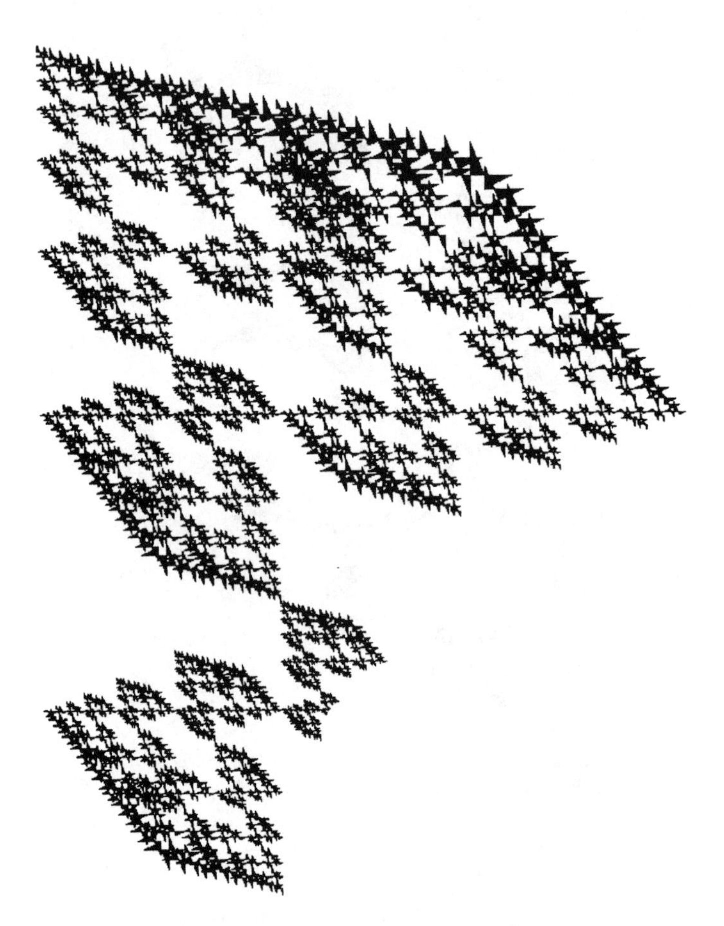

**Figure 7-18**
**Stage 8 of the construction of a fractal with nonright angles measuring 24° and 66°.**

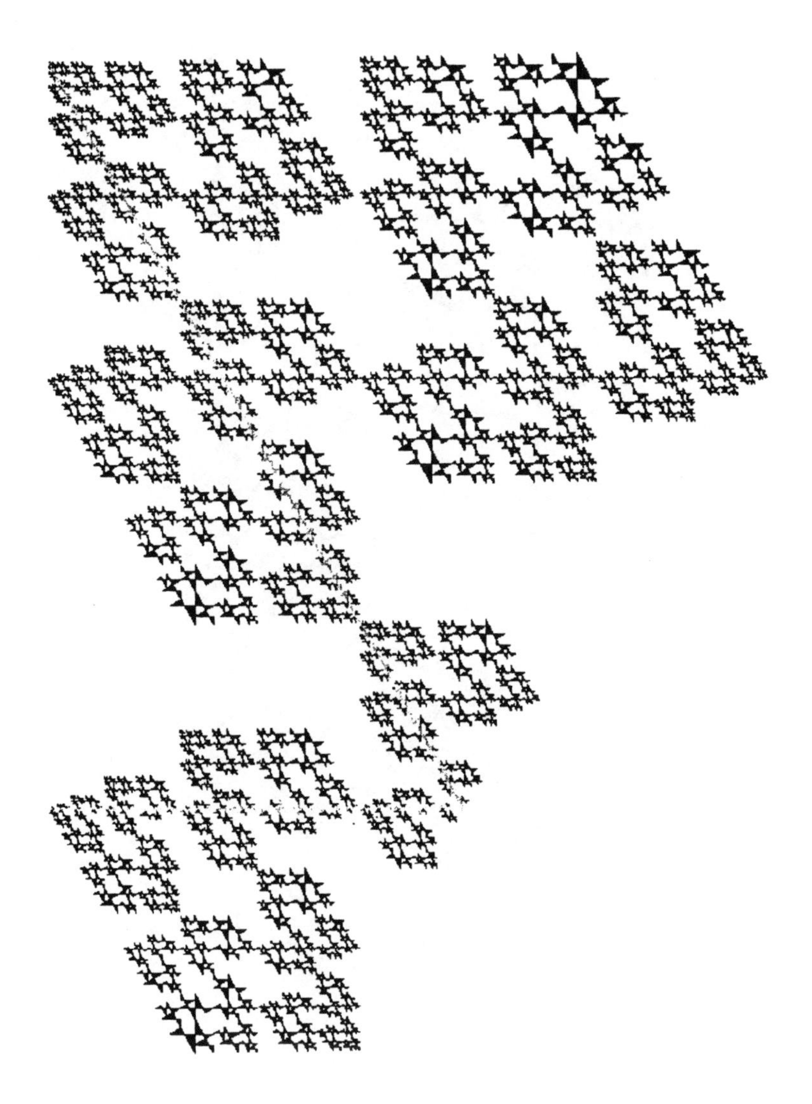

**Figure 7-19**
**Stage 8 of the construction of a fractal with nonright angles measuring 30° and 60°.**

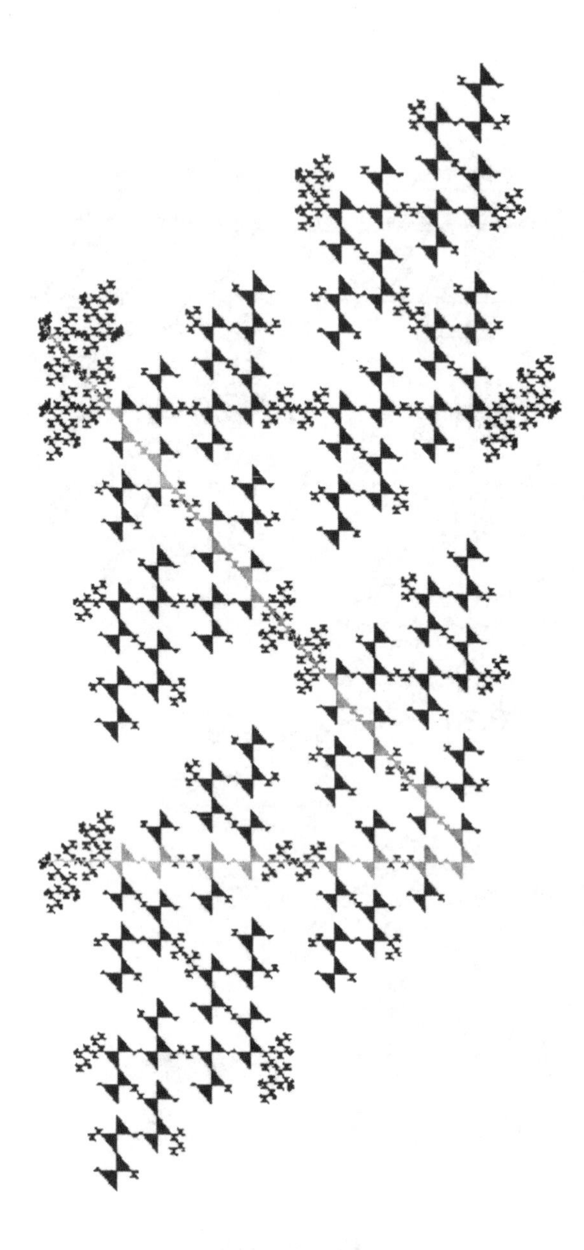

**Figure 7-20**
**Stage 8 of the construction of a fractal with nonright angles measuring 40° and 50°.**

The area and the perimeter of these fractals also show interesting properties. To appreciate what is happening as we repeat, or iterate, the generative procedure with these figures over and over, let us first look closer at what each iteration does.

First let us notice that the procedure shown in figure 7-12 transforms a right triangle into three similar triangles. What are the similarity ratios[3] of the original triangle to the new triangles?

Let us use the labels shown in figure 7-21. Since the triangle of sides $j$, $k$, and $l$ is similar to the original triangle of sides $a$, $b$, and $c$, their corresponding sides are in proportion:

$$\frac{j}{a}=\frac{k}{b}=\frac{l}{c}$$

Analogously, sides $m$, $n$, and $o$ are in proportion to $a$, $b$, and $c$, respectively:

$$\frac{m}{a}=\frac{n}{b}=\frac{o}{c}$$

And so are $p$, $q$, and $r$:

$$\frac{p}{a}=\frac{q}{b}=\frac{r}{c}$$

Using these proportions and the relations shown in figure 7-21, we can find the similarity ratios to be $\frac{b-a}{b}$, $\left(\frac{c}{b}-\frac{a}{c}\right)$, and $\frac{a}{c}$.

3. The *similarity ratio*, or sometimes called the *ratio of similitude*, is the quotient you get when you divide any side of one triangle by its corresponding side in the similar triangle.

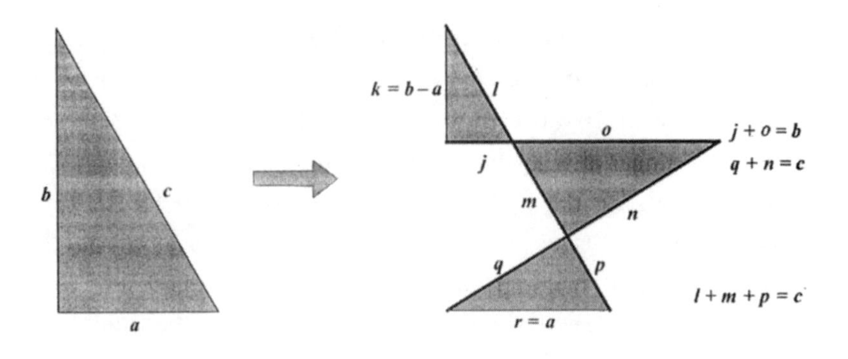


**Figure 7-21**

Specifically, sides $j$, $k$, and $l$ can each be obtained from the sides of the original triangle by multiplying the corresponding side by the ratio $\frac{b-a}{b}$:

$$j = a\left(\frac{b-a}{b}\right)$$

$$k = b\left(\frac{b-a}{b}\right)$$

$$l = c\left(\frac{b-a}{b}\right)$$

Sides $m$, $n$, and $o$ can be obtained from the sides of the original triangle by multiplying corresponding sides by the factor $\left(\frac{c}{b} - \frac{a}{c}\right)$:

$$m = a\left(\frac{c}{b} - \frac{a}{c}\right)$$

$$n = b\left(\frac{c}{b} - \frac{a}{c}\right)$$

$$o = c\left(\frac{c}{b} - \frac{a}{c}\right)$$

Finally $p$, $q$, and $r$ are obtained from $a$, $b$, and $c$ by the ratio $\frac{a}{c}$:

$$p = a\left(\frac{a}{c}\right)$$

$$q = b\left(\frac{a}{c}\right)$$

$$r = c\left(\frac{a}{c}\right) = a$$

And this is what happens at every iteration: each triangle in the figure is replaced by three new triangles, with sides multiplied by the ratios $\frac{b-a}{b}$, $\left(\frac{c}{b}-\frac{a}{c}\right)$, and $\frac{a}{c}$.

Therefore, the perimeter that started out as $a+b+c$, at the next stage is $j+k+l+m+n+o+p+q+r$, or, using the relations we just found:

$$(a+b+c)\left[\frac{b-a}{b}+\left(\frac{c}{b}-\frac{a}{c}\right)+\frac{a}{c}\right]$$

which can be simplified to:

$$(a+b+c)\left(\frac{b-a}{b}+\frac{c}{b}\right) = (a+b+c)\left(\frac{b-a+c}{b}\right)$$

So we see the perimeter is increasing by a factor of $\left(\frac{b-a+c}{b}\right)$.

If we let $P_n$ be the perimeter of the fractal at Stage $n$, we have:

$$P_n = P_{n-1}\left(\frac{b-a+c}{b}\right)$$

That is, the perimeter at a stage is always the perimeter at the previous stage multiplied by $\left(\frac{b-a+c}{b}\right)$.

As an example, let us use as our "seed" the triangle (3, 4, 5).

$$P_0 = a+b+c = 3+4+5 = 12$$

$$P_1 = P_0\left(\frac{b-a+c}{b}\right) = 12\left(\frac{4-3+5}{4}\right) = 12\left(\frac{3}{2}\right) = 18$$

$$P_2 = 18\left(\frac{3}{2}\right) = 27$$

and so forth.

Figure 7-22 shows the values of the perimeter for the first 12 stages of this fractal.

| **Stage (*n*)** | **Perimeter ($P_n$)** |
|---|---|
| 0 | 12 |
| 1 | 18 |
| 2 | 27 |
| 3 | 40.5 |
| 4 | 60.75 |
| 5 | 91.125 |
| 6 | 136.6875 |
| 7 | 205.0313 |
| 8 | 307.5469 |
| 9 | 461.3203 |
| 10 | 691.9805 |
| 11 | 1037.971 |
| 12 | 1556.956 |

**Figure 7-22**

We can see that the perimeter will increase without bound as we perform more iterations.

The fraçtal's area, on the other hand, started out in Stage 0 as $\frac{ab}{2}$ (since side $a$ can be considered as the base and side $b$ as the height), and in Stage 1 became:

$$\frac{jk}{2}+\frac{mn}{2}+\frac{pq}{2}$$

or

$$\left(\frac{ab}{2}\right)\left[\left(\frac{b-a}{b}\right)^2+\left(\frac{c}{b}-\frac{a}{c}\right)^2+\left(\frac{a}{c}\right)^2\right]$$

If we let $A_n$ be the area of the fractal at Stage $n$, we have:

$$A_n = A_{n-1}\left[\left(\frac{b-a}{b}\right)^2+\left(\frac{c}{b}-\frac{a}{c}\right)^2+\left(\frac{a}{c}\right)^2\right]$$

In the case of the fractal that originates from triangle (3, 4, 5), this expression becomes:

$$A_n = A_{n-1}\left[\left(\frac{4-3}{4}\right)^2+\left(\frac{5}{4}-\frac{3}{5}\right)^2+\left(\frac{3}{5}\right)^2\right] = A_{n-1}\times 0.845$$

which shows that the area being multiplied by a factor between 0 and 1 will decrease at each stage and approach zero (figure 7-23).

And so we get at another counterintuitive property in this mathematical object. At its limit, this fractal has an infinite perimeter but an area of zero.

| **Stage ($n$)** | **Area ($A_n$)** |
|---|---|
| 0 | 6 |
| 1 | 5.07 |
| 2 | 4.28415 |
| 3 | 3.620107 |
| 4 | 3.05899 |
| 5 | 2.584847 |
| 6 | 2.184195 |
| 7 | 1.845645 |
| 8 | 1.55957 |
| 9 | 1.317837 |
| 10 | 1.113572 |
| 11 | 0.940968 |
| 12 | 0.795118 |

**Figure 7-23**

In this chapter, we used only two of the configurations you encountered before in this book, and with them we created this gallery of Pythagorean fractals. The most remarkable thing is that the slightest tweak in procedures—a change in the orientation of triangles, or a slight change in angles—escalates when it is repeated ad infinitum, producing results that have totally different overall and sometimes breathtaking appearances.

The various diagrams used elsewhere in this book to demonstrate the Pythagorean Theorem all offer great potential for beautiful constructions. With creativity and access to suitable computer programs, the reader can undertake an adventure to produce objects of amazing appearance and properties. And even if a computer is not available, patience and artistic talent will create fascinating results—as the works of Albert E. Bosman and Jos de Mey attest. The inspiration is in the properties found when we contemplate the squares on the sides of right triangles. These are the same properties that fascinated Pythagoras as well as so many others since antiquity.

# Final Thoughts

Now that we have explored the Pythagorean Theorem from many different vantage points, it is important to realize, above all, that this theorem opened up the study of mathematics to more people than any other in the field. Even though we will probably never be fully certain who in history first discovered the relationships of this theorem, it remains the nominal ownership of Pythagoras.

The beauty captured by this theorem depends on the perception of the viewer. Those who are enchanted by geometric relationships never cease to be amazed at the clever visual justifications of this famous theorem. They seem to motivate the dedicated to further seek other such visual proofs of it. Some of the more dramatic demonstrations have been explored with the hope that the reader will continue to search out other geometric relationships that portray this glorious theorem.

For those who admire numerical relationships, the Pythagorean Theorem provides a plethora of numerical connections to many other seemingly unrelated mathematical theorems, patterns, and formulas. One such pattern is the well-known Fibonacci sequence, which, as we mentioned earlier, is completely independent of the Pythagorean Theorem and yet can be shown to have a connection to it. Once again, we hope to have opened the door for the reader to discover other numerical relationships to the Pythagorean Theorem.

From an artistic standpoint, the Pythagorean Theorem has given rise to a subdivision of fractals that besides having serious mathematical significance, also offers aesthetic pleasure to many. These fractals provide an intriguing form of regimented art that evokes the joy of unexpected visual delights.

The work of the Pythagoreans touched on many fields, both within mathematics and beyond. They pursued measures of "central tendency" of various sorts. These measures of central tendency—or means, as we call them—are an important part of today's quantitative understanding of the world. Throughout these pages, we have gotten to better appreciate these all-important means—arithmetic, geometric, and harmonic—by inspecting both algebraically and geometrically how they interrelate with each other.

Another aspect of the Pythagoreans' work dealt with music. Although a departure from this famous theorem, today's compositions, without the Pythagoreans' contributions to music, might be quite different.

Thus the work of Pythagoras and his followers has expanded our visual, quantitative, intellectual, and auditory perceptions. It is hoped that all have gained a better appreciation of these wonders as we pursued the power and the beauty of the Pythagorean Theorem.

## Afterword

# About the Mathematics Work That Led to the 1985 Nobel Prize in Chemistry: Thanks Ultimately to Pythagoras

**by Dr. Herbert A. Hauptman**

A particularly clear example of the important role that mathematics plays in promoting the growth of science and technology in the twenty-first century is provided by the work that led to the 1985 Nobel Prize in Chemistry: the solution of the so-called phase problem of x-ray crystallography. This example makes transparent, in the clearest way possible, the interplay between science and mathematics.

When a beam of x-rays strikes a crystal, the incident beam is split into many weaker beams having different directions and different intensities, thus giving rise to what's called the diffraction pattern. The nature of the diffraction pattern, which is to say the directions and intensities of the scattered x-rays, is determined by the structure of the crystal—that is, the arrangement of the atoms in the crystal. If one knew the structure of the crystal, one could readily predict the nature of the diffraction pattern.

However, the problem facing the crystallographer is the converse: One observes the diffraction pattern—that is, one measures the directions and intensities of the x-rays scattered by the crystal. Can one then deduce the structure of the crystal—that is, the atomic arrangement—that gives rise to the observed diffraction pattern? The 1985 Nobel Prize in Chemistry was given for the solution of this problem.

The ability to identify crystal structures rapidly and routinely had important consequences. Possibly the most important was the ability to relate crystal and molecular structures to biological activity. Thus it became possible to understand life processes at the "molecular" level, to understand better how living things "work" and the cause of disease, and to devise better therapies and better drugs for the prevention and treatment of disease with a minimum of adverse side effects—in short, to improve human health.

Finally, we would be remiss if we did not give due credit to the Greek mathematician Pythagoras, who made the fundamental contribution that is arguably more often cited than any other. This is simply the relationship $a^2 + b^2 = c^2$, where $a$ and $b$ are the lengths of the legs of the right triangle of which $c$ is the hypotenuse. Without this relationship the whole science of x-ray crystallography would not exist as it does today. If the science of x-ray crystallography had not been developed in the nineteenth and twentieth centuries as it had been, our ability to understand biological processes would have been compromised. Thus, the argument can plausibly be made that we in the Western world owe our high standard of medical care in large part to the Greek mathematician Pythagoras.

The relevance of the science of x-ray crystallography in the twentieth century suggests that one should first distinguish between science and technology.

Science is the discipline that attempts to describe the reality of the world around us, including the nature of living organisms, by rational means. Technology, on the other hand, attempts to exploit the fruits of science in order to attain human goals. In short, science is to be thought of as related to knowledge, while technology is concerned with the utilization of knowledge for, one hopes, the betterment of the human condition.

Four hundred years ago, no one could possibly have anticipated the enormous strides that science and technology were destined to make in the ensuing centuries. Even as recently as one hundred years ago, who would have predicted the great revolutions in these two disciplines that the twentieth century held in store for us? Thus the theories of relativity and quantum mechanics, the nature of the structure of matter, molecular biology, and our new understanding of life processes changed forever the way we look at the world around us, and at the same time they have irrevocably established the rational mode of inquiry, the quintessential element of the scientific method, as preferred above all others.

Technological applications, as we'll see a mixed blessing at best, followed quickly on the heels of the more basic scientific discoveries; the fruits of technology then fed back into and facilitated the increasingly rapid advance of science, so that today we are racing ahead at breakneck speed to a future filled with uncertainty. Among the notable accomplishments of the past century were the invention and rapid development of the digital computer; our remarkable progress in communications, transportation, space exploration, and electronics; improved methods for the diagnosis and treatment of disease; and the use of the atom as the source of limitless amounts of energy. On the dark side, however, were the development and perfection of intercontinental missiles armed with nuclear warheads; atomic, chemical, and biological means of mass destruction; and the pollution of the environment. These were only a few of the consequences—not all inevitable—of the scientific revolutions of the twentieth century.

Thus, it is clear that the spectacular advances of science and technology in the twentieth century and the current trends hold enormous promise for good, and an equally great threat to our very survival. The promise is that the fruits of science will be used for the benefit and well-being of humankind, leading to a never-ending improvement in the quality of life for everyone; the threat is that the fruits of science will be used for destructive purposes, leading to consequences ranging from devastating pollution of the environment to the destruction of human life by nuclear holocaust.

The threat arises from the crisis produced by the lightning advances of science and technology on the one hand and the glacial evolution of mental attitudes and modes of behavior, measured in periods of centuries, on the other hand. This conflict, between science and conscience, between technology and ethical behavior, has now reached the point where it threatens the destruction of humankind, if not of the planet itself, unless it is resolved on favorable terms, and resolved soon.

The problem is made more difficult by the fact that one and the same scientific discovery may find application in different ways, some good and others not so good. Thus, the energy of the atom may be used to generate useful power, to treat disease, or to destroy life. So what is to be done? I don't know the answer to this question, but in making these remarks I want simply to stress that this dilemma exists, and is one in a sense which scientists have created but which all of humanity must resolve. The solution will tax the ingenuity of all.

# Pictorial Depictions of Pythagoras and His Famous Theorem

No one really knows what Pythagoras looked like, yet we have a variety of pictures that claim to have his image. The following postage stamps will give you some idea how he has been perceived. You can then draw your own conclusions about his appearance. We also show some stamps that further popularize his most famous mathematics theorem.

## Stamps with a Depiction of Pythagoras

**Greece, 1955**

**Sierra Leone, 1983**

**Enlargement of Pythagoras from above**

**From Raphael's *The School of Athens* (Raffaello Sanzio, 1511)**

**San Marino, 1983**

## Further Depictions of Pythagoras

## The Pythagorean Theorem on Stamps

Macedonia, 1998

**Japan, 1984**

**Sierra Leone, 1984**

**Nicaragua, 1971**

**Greece, 1955**

**Greece, 1955**

**Italy**

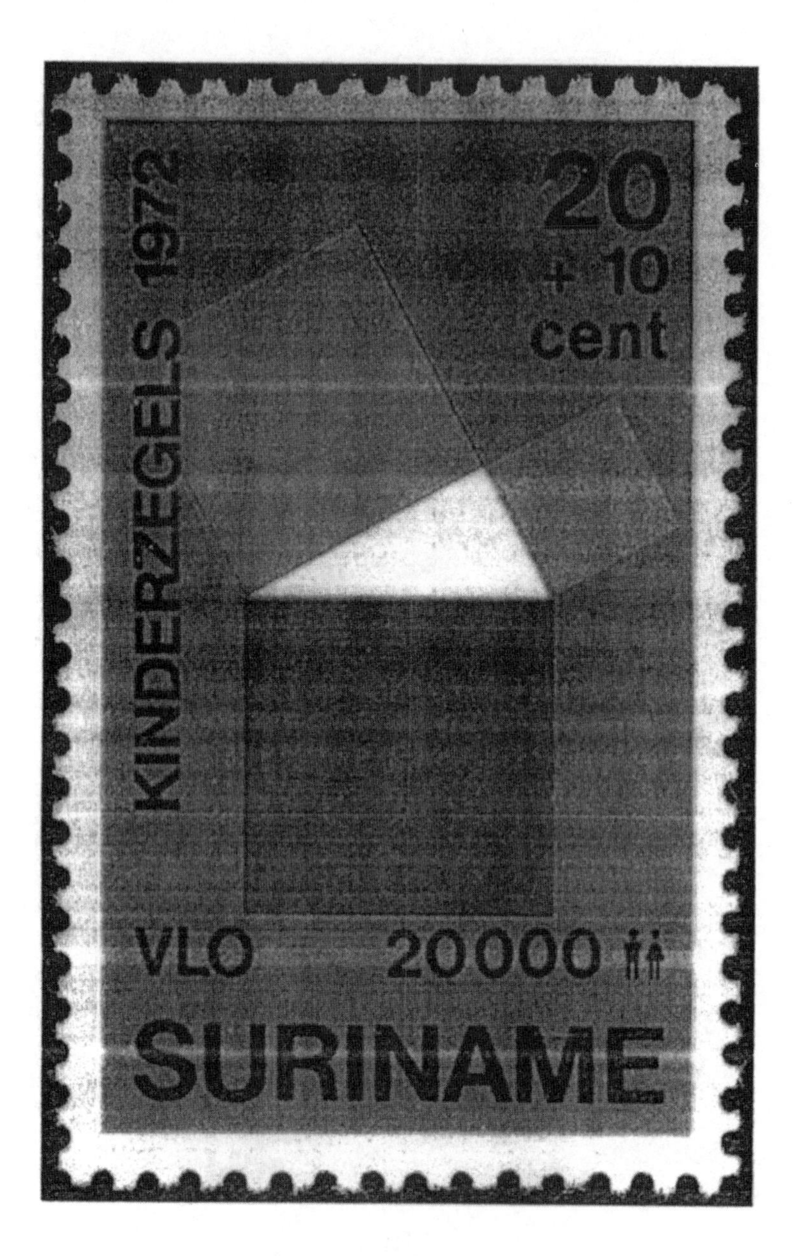


**Republic of Suriname, 1972**

## Fermat's Extension of the Pythagorean Theorem



**Czech Republic, 2000**

**France**

# Appendix A

# Some Selected Proofs

## Determining if an angle is acute or obtuse (page 81)

Consider $\Delta ABC$, in which $c$ is the measure of the longest side and $\angle C$ has the greatest measure. We must consider two cases: $\angle C$ obtuse (figure A-1) and $\angle C$ acute (figure A-2). Since the two cases are similar, we shall prove them at the same time.

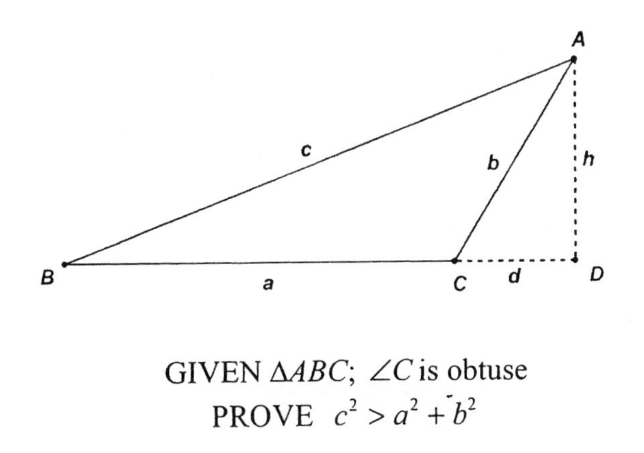

**Figure A-1**

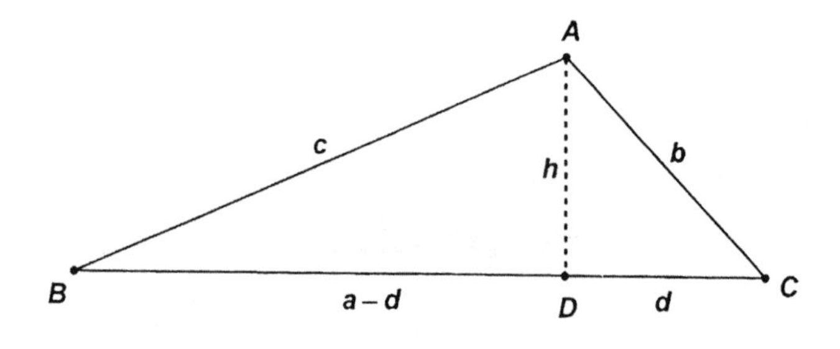


GIVEN $\Delta ABC$; $\angle C$ is acute
PROVE $c^2 < a^2 + b^2$

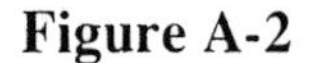
**Figure A-2**

## PROOF

1. $\overline{AD}$ is an altitude of $\Delta ABC$. Let $DC = d$.

| **For $\angle C$ obtuse** | **For $\angle C$ acute** |
|---|---|
| 2. $BD = a + d$ | 2. $BD = a - d$ |

Apply the Pythagorean Theorem to $\Delta ABD$. Let $AD = h$.

| | |
|---|---|
| 3. $c^2 = h^2 + (a + d)^2$ | 3. $c^2 = h^2 + (a - d)^2$ |
| 4. $c^2 = h^2 + a^2 + d^2 + 2ad$ | 4. $c^2 = h^2 + a^2 + d^2 - 2ad$ |

5. Apply the Pythagorean Theorem to $\Delta ACD$; $b^2 = h^2 + d^2$ in both cases.

Substitute the result in step 5 into the equations in step 4:

| | |
|---|---|
| 6. $c^2 = a^2 + b^2 + 2ad$ | 6. $c^2 = a^2 + b^2 - 2ad$ |
| 7. $c^2 > a^2 + b^2$ | 7. $c^2 < a^2 + b^2$ |

We have now established that in an obtuse triangle, the square of the length of the longest side is greater than the sum of the squares of the lengths of the two shorter sides.

We have also established that in an acute triangle, the square of the length of the longest side is less than the sum of the squares of the lengths of the two shorter sides.

The converse, which also can be proved true, tells us that if $a^2 + b^2 > c^2$, then the angle is acute and if $a^2 + b^2 < c^2$, then the angle is obtuse.

## Pythagorean extensions (page 104)

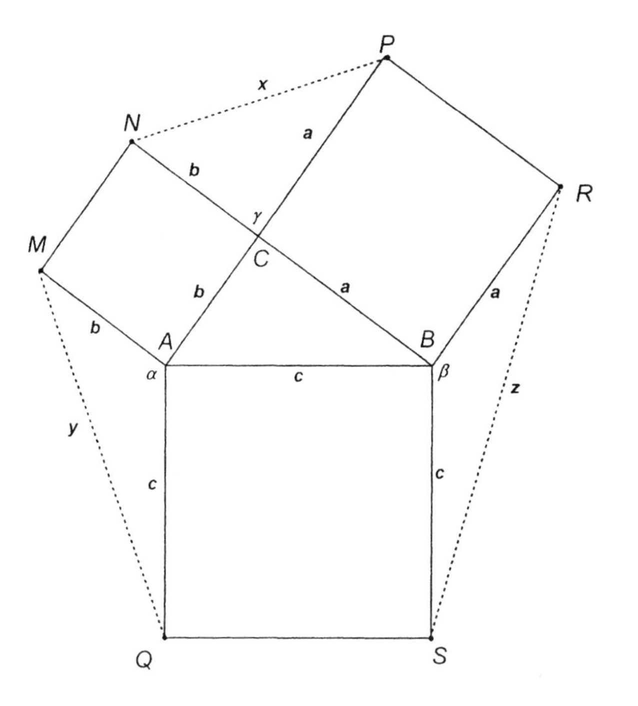

**Figure A-3**

We begin with any triangle $ABC$ with squares on its sides.

**To prove** $x^2 + y^2 + z^2 = 3(a^2 + b^2 + c^2)$, we apply the law of cosines to the three triangles (figure A-3):

$$y^2 = b^2 + c^2 - 2bc\cos\alpha$$
$$x^2 = a^2 + b^2 - 2ab\cos\gamma$$
$$z^2 = a^2 + c^2 - 2ac\cos\beta$$

We then add these three equations, getting

$$x^2 + y^2 + z^2 = 2(a^2 + b^2 + c^2) - 2(bc\cos\alpha + ab\cos\gamma + ac\cos\beta) \qquad \text{(I)}$$

We then apply the law of cosines to $\Delta ABC$

$$a^2 = b^2 + c^2 - 2bc\cos A$$
$$b^2 = a^2 + c^2 - 2ac\cos B$$
$$c^2 = b^2 + a^2 - 2ba\cos C$$

where ($A$, $B$, and $C$ are the interior angles of $\Delta ABC$). The addition of these three equations gives us

$$a^2 + b^2 + c^2 = 2(a^2 + b^2 + c^2) - 2(bc\cos A + ac\cos B + ab\cos C),$$

or

$$2(bc\cos A + ac\cos B + ab\cos C) = a^2 + b^2 + c^2 \qquad \text{(II)}$$

Since the squares on the sides $a$, $b$, $c$ form right angles $A$, $B$, $C$, we get $\alpha = 180^\circ - A$, $\gamma = 180^\circ - C$, and $\beta = 180^\circ - B$. Substituting in equation (I) and using identity $\cos(180^\circ - \theta) = -\cos\theta$, gives us

$$a^2 + b^2 + c^2 = 2(a^2 + b^2 + c^2) + 2(bc\cos A + ac\cos B + ab\cos C) \qquad \text{(III)}$$

Substituting equation (II) into equation (III) gives us

$$\begin{aligned} x^2 + y^2 + z^2 &= 2(a^2 + b^2 + c^2) + (a^2 + b^2 + c^2) \\ &= 3(a^2 + b^2 + c^2) \end{aligned}$$

## Proof that for parallelogram *ABCD* (figure A-4) $m^2 + n^2 = a^2 + b^2 + c^2 + d^2$ (page 110)

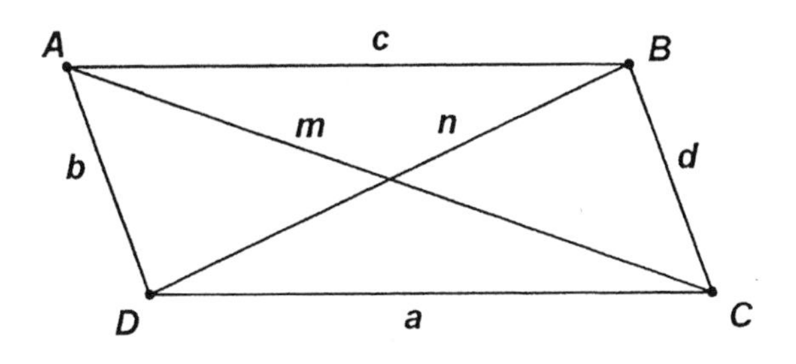


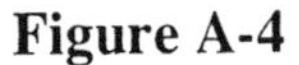
**Figure A-4**

Let $\angle ADC = \alpha$. Then $\angle DAB = 180° - \alpha$ (see figure A-4). By the law of cosines, we get

$$\Delta ACD: \; m^2 = a^2 + b^2 - 2ab\cos\alpha$$

$$\Delta ABD: n^2 = c^2 + b^2 - 2cb\cos(180° - \alpha) = c^2 + b^2 + 2ab\cos\alpha$$

since $b = d$ and $a = c$.

With the equality of the pairs of opposite sides of the parallelogram, adding the two equations gives us

$$m^2 + n^2 = a^2 + b^2 + c^2 + d^2$$

## Proof that the length of the median drawn to the hypotenuse of a right triangle is half the length of the hypotenuse (page 95)

In figure A-5, *CD* is the median of triangle *ABC* drawn to hypotenuse *AB*. We draw perpendiculars *DE* and *DF* to sides *AC* and *BC*, respectively, determining points *E* and *F* as midpoints of their respective sides. This makes triangles *ADC* and *BDC* isosceles and, hence, *AD* = *DC* = *DB*. We can also circumscribe a circle about right $\Delta ABC$, since $\angle C = 90°$. Side $AB$ is the circle's diameter and then $\frac{1}{2}AB = DB = CD$, as they are radii of the circle.

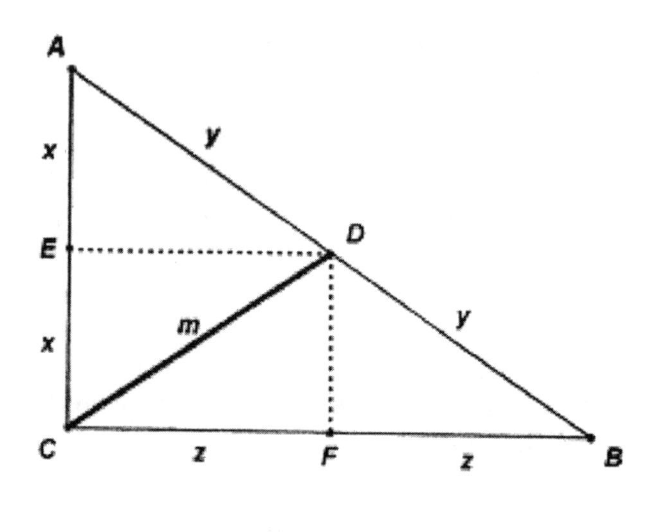


**Figure A-5**

## Proof of Fibonacci generation of Pythagorean triples (page 137)

Suppose $a$, $b$, $c$, $d$ form a Fibonacci sequence.
Then we get $c = a + b$, $d = c + b = a + b + b = a + 2b$.

The operations indicated in the procedure are

$$A = \mathbf{2bc} = 2b(a+b) = 2ab + 2b^2$$
$$B = \mathbf{ad} = a(a+2b) = a^2 + 2ab$$
$$C = \mathbf{b^2} + \mathbf{c^2} = b^2 + (a+b)^2 = a^2 + 2ab + 2b^2$$

Applying the Pythagorean theorem to see if these triples fit the theorem:

$$A^2 = (2ab + 2b^2)^2 = 4a^2b^2 + 8ab^3 + 4b^4$$
$$B^2 = (a^2 + 2ab)^2 = a^4 + 4a^3b + 4a^2b^2$$
$$C^2 = (a^2 + 2ab + 2b^2)^2 = a^4 + 4a^3b + 8a^2b^2 + 8ab^3 + 4b^4$$

$$\begin{array}{ccccc} A^2 & + & B^2 & = & C^2 \end{array}$$
$$[4a^2b^2 + 8ab^3 + 4b^4] + [a^4 + 4a^3b + 4a^2b^2] = [a^4 + 4a^3b + 8a^2b^2 + 8ab^3 + 4b^4]$$

First we'll prove the following lemma, which will come in handy here and later on.

**Lemma.** $F_{m+n} = F_{m-1}F_n + F_mF_{n+1}$

*Proof of Lemma:*

We proceed by induction on $n$. (Actually, this will be a form of induction that is called "strong induction," where we will be using the statement for $n = k - 1$ and for $n = k$ to prove the statement for $n = k + 1$. This will also mean that we'll need two base cases rather than just one; so we'll have to check the statement for $n = 1$ and for $n = 2$.) For $n = 1$, we have to check that $F_{m+1} = F_{m-1}F_1 + F_mF_2$, or, since $F_1 = 1$ and $F_2 = 1$, we have to check that $F_{m+1} = F_{m-1} + F_m$; and this is of course true because it's the very relation by which we define the Fibonacci numbers. For $n = 2$, we have to check that $F_{m+2} = F_{m-1}F_2 + F_mF_3$, or, since $F_2 = 1$ and $F_3 = 2$, we have to check that $F_{m+2} = F_{m-1} + 2F_m$; and this is true by the following sequence of equalities:

$$F_{m-1} + 2F_m = (F_{m-1} + F_m) + F_m = F_{m+1} + F_m = F_{m+2}$$

Now assume the statement is true for $n = k - 1$ and $n = k$, that is, assume

$$F_{m+k-1} = F_{m-1}F_{k-1} + F_mF_k$$
and
$$F_{m+k} = F_{m-1}F_k + F_mF_{k+1}$$

(This is our induction hypothesis.) Then

$$\begin{aligned}
&F_{m-1}F_{k+1} + F_mF_{k+2} \\
&= F_{m-1}(F_{k-1} + F_k) + F_m(F_k + F_{k+1}) \\
&= F_{m-1}F_{k-1} + F_{m-1}F_k + F_mF_k + F_mF_{k+1} \\
&= F_{m-1}F_k + F_mF_{k+1} + F_{m-1}F_{k-1} + F_mF_k \\
&= F_{m+k} + F_{m+k-1} \\
&= F_{m+k+1}
\end{aligned}$$

That is, $F_{m-1}F_{k+1} + F_mF_{k+2} = F_{m+k+1}$, which is exactly the statement for $n = k + 1$. So our induction is complete.

We are now ready to prove the relationship $F_n^2 + F_{n+1}^2 = F_{2n+1}$; that is, the sum of the squares of the Fibonacci numbers in positions $n$ and $n + 1$ (consecutive positions) is the Fibonacci number in place $2n + 1$.

*Proof:*

In the lemma, let $m = n + 1$. Then we get $F_{2n+1} = F_nF_n + F_{n+1}F_{n+1}$, or, in other words, $F_{2n+1} = F_n^2 + F_{n+1}^2$, which is the equation we wanted.

## Proving Ptolemy's Theorem (page 62)

*The product of the lengths of the diagonals of a cyclic quadrilateral equals the sum of the products of the lengths of the pairs of opposite sides. (Ptolemy's Theorem)*

Two methods for proving Ptolemy's Theorem are provided. The second method incorporates the proof of the converse as well, which we state.

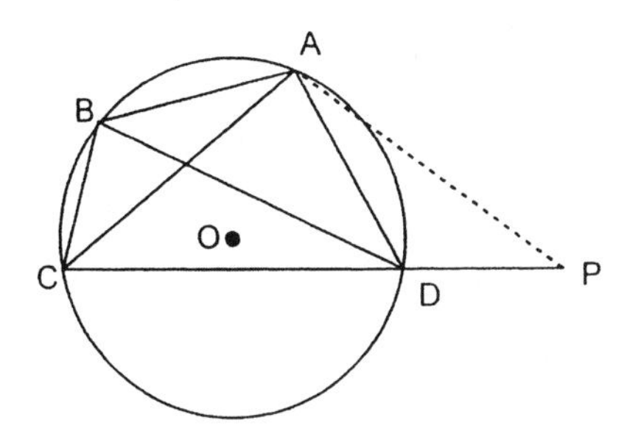


**Figure A-6**

*Proof I*:

In figure A-6, quadrilateral $ABCD$ is inscribed in circle $O$. A line is drawn through $A$ to meet $CD$ at $P$, so that $\angle BAC = \angle DAP$. (I)

Since quadrilateral $ABCD$ is cyclic, $\angle ABC$ is supplementary to $\angle ADC$. However, $\angle ADP$ is also supplementary to $\angle ADC$.

Therefore, $\angle ABC = \angle ADP$ (II)

Thus, $\Delta BAC \sim \Delta DAP$ (III)

and $\frac{AB}{AD} = \frac{BC}{DP}$, or $DP = \frac{(AD)(BC)}{AB}$ (IV)

From (I), $\angle BAD = \angle CAP$, and from (III), $\frac{AB}{AD} = \frac{AC}{AP}$

Therefore, $\Delta ABD \sim \Delta ACP$ and $\frac{BD}{CP} = \frac{AB}{AC}$, or

$$CP = \frac{(AC)(BD)}{AB} \quad \text{(V)}$$

$$CP = CD + DP \qquad \text{(VI)}$$

Substituting (IV) and (V) into (VI), we get

$$\frac{(AC)(BD)}{AB} = CD + \frac{(AD)(BC)}{AB}$$

Thus, $(AC)(BD) = (AB)(CD) + (AD)(BC)$

*Proof II*:

In quadrilateral $ABCD$ (figure A-7), draw $\Delta DAP$ on side $\overline{AD}$ similar to $\Delta CAB$.

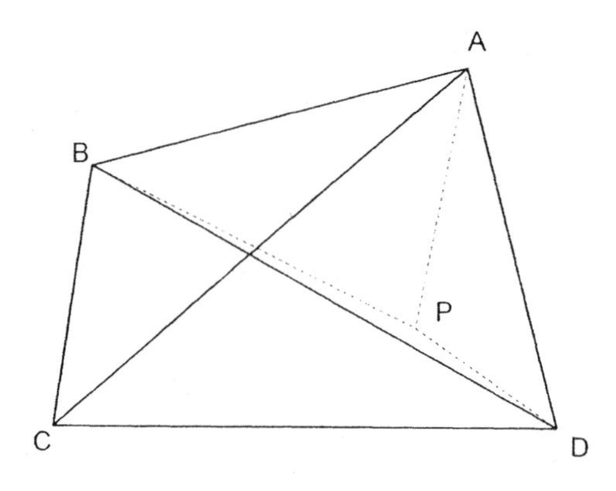

**Figure A-7**

Thus, $\dfrac{AB}{AP} = \dfrac{AC}{AD} = \dfrac{BC}{PD}$ (I)

and $(AC)(PD) = (AD)(BC)$ (II)

Since $\angle BAC = \angle PAD$,

then $\angle BAP = \angle CAD$

Therefore, from (I), $\Delta BAP \sim \Delta CAD$ and $\dfrac{AB}{AC} = \dfrac{BP}{CD}$

or $(AC)(BP) = (AB)(CD)$ (III)

Adding (II) and (III), we have

$$(AC)(BP+PD) = (AD)(BC) + (AB)(CD) \qquad \text{(IV)}$$

Now $BP + PD > BD$ (triangle inequality), unless $P$ is on $\overline{BD}$. However, $P$ will be on $\overline{BD}$ if and only if $\angle ADP = \angle ADB$. But we already know that $\angle ADP = \angle ACB$ (similar triangles). And if $ABCD$ were cyclic, then $\angle ADB$ would equal $\angle ACB$ and $\angle ADB$ would equal $\angle ADP$. Therefore, we can state that if and only if $ABCD$ is cyclic, $P$ lies on $\overline{BD}$.

This tells us that $BP + PD = BD$. (V)

By substituting (V) into (IV),

$$(AC)(BD) = (AD)(BC) + (AB)(CD)$$

Notice we have proved Ptolemy's Theorem *and* its *converse*.

# Appendix B

# Some More Proofs and Solutions

## Comparison of means for *n* items (page 180)

To show that AM $\geq$ GM, recall the definition of the geometric mean

$$g = \sqrt[n]{a_1 \cdot a_2 \cdot a_3 \cdot \ldots \cdot a_n}$$

(where $a_i > 0$)

$$\text{Then, } 1 = \sqrt[n]{\frac{a_1}{g} \cdot \frac{a_2}{g} \cdot \frac{a_3}{g} \cdot \ldots \cdot \frac{a_n}{g}}$$

$$\text{Therefore, } 1 = \frac{a_1}{g} \cdot \frac{a_2}{g} \cdot \frac{a_3}{g} \cdot \ldots \cdot \frac{a_n}{g}$$

We can prove[1] that if the product of $n$ positive numbers equals 1, their sum is *not* less than. Thus we can conclude that

$$\frac{a_1}{g}+\frac{a_2}{g}+\frac{a_3}{g}+\ldots+\frac{a_n}{g}\geq n$$

---

1. To prove: If $x_1, x_2, \ldots, x_n$ are positive and $x_1x_2\ldots x_n = 1$,

then $x_1 + x_2 + \ldots + x_n \geq n$.

First, note that if the $x_i$ are all equal, then each of them must be equal to 1 so that the sum is exactly $n$. In what follows, we will assume that the $x_i$ are not all equal and will show that the sum is strictly greater than $n$. We proceed by induction.

We start with $n = 2$, since, in the case $n = 1$, there is nothing to prove. So assume $x_1x_2 = 1$.

Since $x_1 \neq x_2$, $\sqrt{x_1} - \sqrt{x_2} \neq 0$.

so $0<\left(\sqrt{x_1}-\sqrt{x_2}\right)^2=\left(\sqrt{x_1}\right)^2+\left(\sqrt{x_2}\right)^2-2\sqrt{x_1}\sqrt{x_2}=x_1+x_2-2\sqrt{x_1x_2}$.

However, the radicand equals 1, so the inequality then reduces to $0<x_1+x_2-2$ or $x_1+x_2>2$ as required.

Our induction hypothesis for $n = k$ is that for any positive $y_1, y_2, \ldots, y_k$, not all equal, such that $y_1y_2\ldots y_k=1$ we have $y_1+y_2+\ldots+y_k>k$.

Now suppose that $x_1x_2\ldots x_{k+1} = 1$. If all the $x_i$ were greater than 1, then the product would be too. Likewise, if all the $x_i$ were less than 1 then so would their product. Therefore, it follows that at least one of the $x_i$'s is less than 1 and another is greater than 1. Reordering the $x_i$'s, we may assume that $x_1<1$ and $x_2>1$.

But then $1-x_1>0$ and $1-x_2<0$, so their product is negative; i.e., $0>(1-x_1)\cdot(1-x_2)=1-(x_1+x_2)+x_1x_2$ or $(x_1+x_2)-1>x_1x_2$.

Returning to our assumption on the product of the $x_i$, we have $(x_1x_2)\ldots x_{k+1} = 1$ where we have inserted parentheses to present the product as made up of $k$ (rather than $k + 1$) factors. By our induction hypothesis for $k$ factors, we have that $(x_1x_2)+\ldots+x_{k+1}>k$.

Recalling that $(x_1+x_2)-1>x_1x_2$, we have

$$(x_1+x_2-1)+\ldots+x_{k+1}>x_1x_2+\ldots+x_{k+1}>k$$

Adding 1 to both extremes of the inequality, we get $x_1+x_2+\ldots+x_{k+1}>k+1$, completing the induction.

Therefore, $\frac{a_1 + a_2 + a_3 + \ldots + a_n}{n} \geq g$. This leads us to conclude that $\frac{a_1 + a_2 + a_3 + \ldots + a_n}{n} \geq \sqrt[n]{a_1 \cdot a_2 \cdot a_3 \cdot \ldots \cdot a_n}$, or AM $\geq$ GM.

We will now show that GM $\geq$ HM. Begin by considering the sequence $a_1^b, a_2^b, a_3^b, \ldots, a_n^b$, where $b$ is integral. Since we just proved that AM $\geq$ GM, we can say that

$$\frac{a_1^b + a_2^b + a_3^b + \ldots + a_n^b}{n} \geq \sqrt[n]{a_1^b \cdot a_2^b \cdot a_3^b \cdot \ldots \cdot a_n^b}$$

When $\frac{1}{b} < 0$,

$$\left[\sqrt[n]{a_1^b \cdot a_2^b \cdot a_3^b \cdot \ldots \cdot a_n^b}\right]^{\frac{1}{b}} \geq \left[\frac{a_1^b + a_2^b + a_3^b + \ldots + a_n^b}{n}\right]^{\frac{1}{b}}$$

If we take $b = -1$, then we get

$$\sqrt[n]{a_1^b \cdot a_2^b \cdot a_3^b \cdot \ldots \cdot a_n^b} \geq \left[\frac{a_1^{-1} + a_2^{-1} + a_3^{-1} + \ldots + a_n^{-1}}{n}\right]^{-1}$$

Hence, $\sqrt[n]{a_n \cdot a_n \cdot a_n \cdot \ldots \cdot a_n} \geq \dfrac{n}{\frac{1}{a_1} + \frac{1}{a_2} + \frac{1}{a_3} + \ldots + \frac{1}{a_n}}$,

or GM $\geq$ HM

## The Solutions to the "Nuggets"

*For Nugget 1 (page 182):*

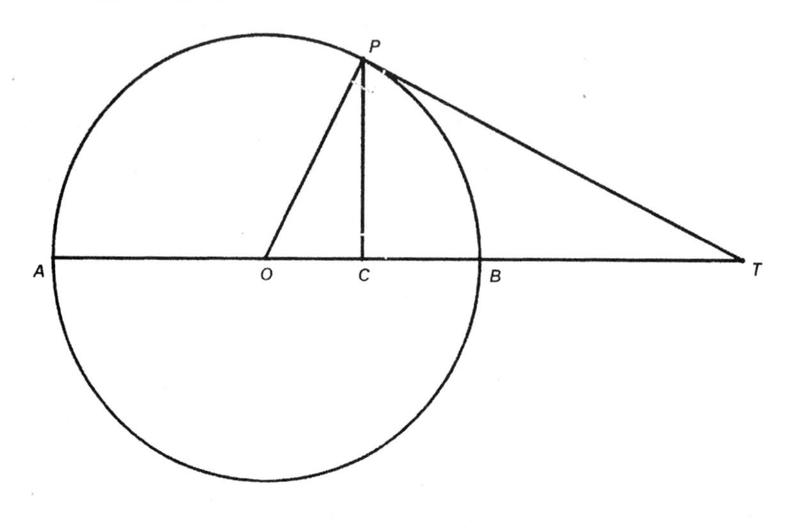


**Figure B-1**

a. In figure B-1, where $AT = a$ and $BT = b$, we will show that $TO$ is the arithmetic mean between $a$ and $b$:

$$a + b = AT + BT$$
$$a + b = AO + OB + BT + BT$$

Since $AO = OB$, $a + b = 2OB + 2BT$.
It follows that $\frac{a+b}{2} = OB + BT = TO$, the arithmetic mean.

b. To show that $PT$ is the geometric mean between $a$ and $b$, we begin with $\Delta APT \sim \Delta PBT$.
Therefore, $\frac{AT}{PT} = \frac{PT}{BT}$,
from which we get $PT = \sqrt{(AT)(BT)} = \sqrt{ab}$, the geometric mean.

c. We can also show that $TC$ is the harmonic mean between $a$ and $b$. In right $\Delta OPT$, $PT$ is the mean proportional between $TC$ and $TO$.

Therefore, $PT^2 = (TC)(TO)$, or $TC = \frac{PT^2}{TO}$.

Since $PT^2 = ab$ and $TO = \frac{a+b}{2}$, we get $TC = \frac{ab}{\frac{1}{2}(a+b)}$.

*For Nugget 2 (pages 182–83):*

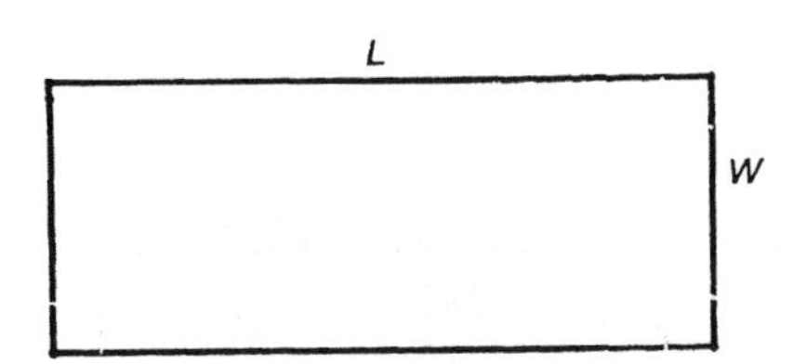

**Figure B-2**

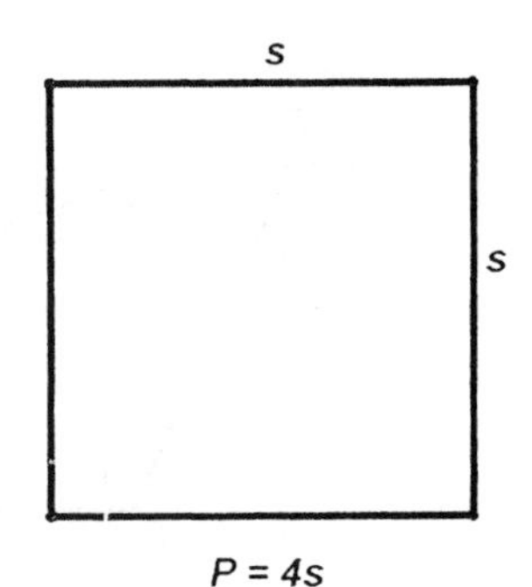


**Figure B-3**

a. The rectangle and the square have the same perimeter as shown in figures B-2 and B-3: $4s = 2L + 2W$.

Then the side of the square is $s = \frac{L+W}{2}$, the arithmetic mean between the sides of the rectangle.

b. The rectangle and the square have the same area. Therefore, $s^2 = LW$, and it follows that $s = \sqrt{LW}$, which is the geometric mean between the sides of the rectangle.

c. The rectangle and the square have the same ratio of area to perimeter, which can be shown as $\frac{LW}{2(L+W)} = \frac{s^2}{4s}$, which then reduces to $\frac{LW}{L+W} = \frac{s}{2}$. Therefore, $s = \frac{2LW}{L+W}$, which is the harmonic mean between the sides of the rectangle.

*For Nugget 3 (page 183):*

The numbers for a cube are: vertices = 8; edges = 12; faces = 6. The harmonic mean between 12 and 6 is $\frac{(2)(12)(6)}{12+6} = 8$.

*For Nugget 4 (page 183):*

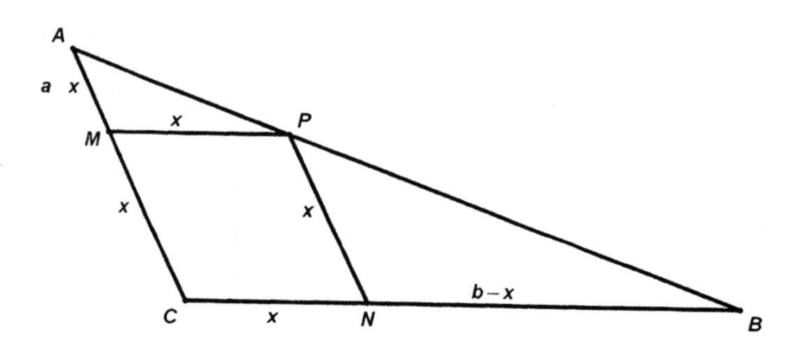


**Figure B-4**

From figure B-4 we begin with the $\Delta AMP \sim \Delta ACB$, which then gives us $\frac{AM}{AC} = \frac{MP}{CB}$ or $\frac{a-x}{a} = \frac{x}{b}$.

Then it follows that $x = \frac{ab}{a+b}$.

Therefore, $MP + NP = 2x = \frac{2ab}{a+b}$, which is the harmonic mean between $a$ and $b$, or $AC$ and $BC$.

*For Nugget 5 (pages 183–84):*

We can establish that in figure B-5 triangles $APQ$ and $ABC$ are similar, which then allows us to establish that $\frac{h-x}{x} = \frac{h}{a}$, and then $x = \frac{ah}{a+h}$. The semiperimeter of the square is $2x$, and $2x = \frac{2ah}{a+h}$, which is the harmonic mean between $AH$ and $BC$.

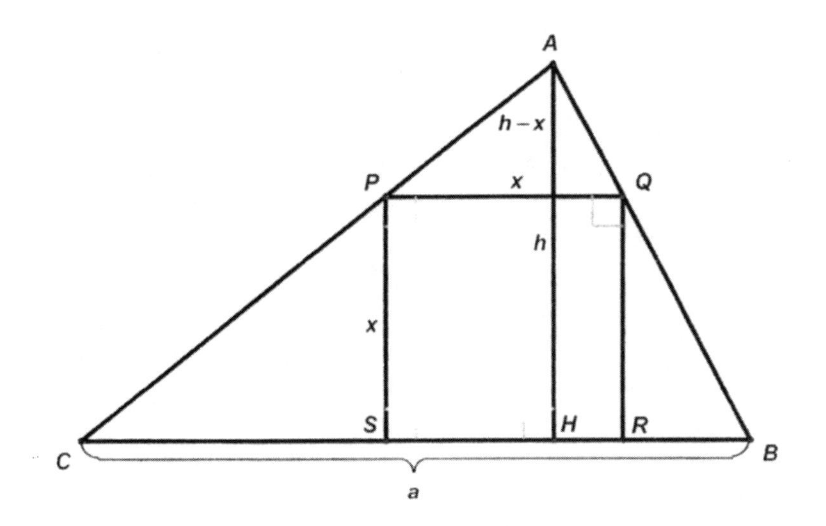


**Figure B-5**

# Appendix C

# List of Primitive Pythagorean Triples

| $m$ | $n$ | $a = m^2 - n^2$ | $b = 2mn$ | $c = m^2 + n^2$ | Pythagorean Triple | Perimeter | Area | In-radius |
|---|---|---|---|---|---|---|---|---|
| 2 | 1 | 3 | 4 | 5 | (3, 4, 5) | 12 | 6 | 1 |
| 3 | 2 | 5 | 12 | 13 | (5, 12, 13) | 30 | 30 | 2 |
| 4 | 1 | 15 | 8 | 17 | (8, 15, 17) | 40 | 60 | 3 |
| 4 | 3 | 7 | 24 | 25 | (7, 24, 25) | 56 | 84 | 3 |
| 5 | 2 | 21 | 20 | 29 | (20, 21, 29) | 70 | 210 | 6 |
| 6 | 1 | 35 | 12 | 37 | (12, 35, 37) | 84 | 210 | 5 |
| 5 | 4 | 9 | 40 | 41 | (9, 40, 41) | 90 | 180 | 4 |
| 7 | 2 | 45 | 28 | 53 | (28, 45, 53) | 126 | 630 | 10 |
| 6 | 5 | 11 | 60 | 61 | (11,60,61) | 132 | 330 | 5 |
| 8 | 1 | 63 | 16 | 65 | (16, 63, 65) | 144 | 504 | 7 |
| 7 | 4 | 33 | 56 | 65 | (33, 56, 65) | 154 | 924 | 12 |
| 8 | 3 | 55 | 48 | 73 | (48, 55, 73) | 176 | 1320 | 15 |
| 7 | 6 | 13 | 84 | 85 | (13, 84, 85) | 182 | 546 | 6 |
| 9 | 2 | 77 | 36 | 85 | (36, 77, 85) | 198 | 1386 | 14 |
| 8 | 5 | 39 | 80 | 89 | (39, 80, 89) | 208 | 1560 | 15 |
| 10 | 1 | 99 | 20 | 101 | (20, 99 101) | 220 | 990 | 9 |
| 9 | 4 | 65 | 72 | 97 | (65, 72, 97) | 234 | 2340 | 20 |
| 8 | 7 | 15 | 112 | 113 | (15, 112 113) | 240 | 840 | 7 |
| 10 | 3 | 91 | 60 | 109 | (60, 91, 109) | 260 | 2730 | 21 |
| 11 | 2 | 117 | 44 | 125 | (44, 117, 125) | 286 | 2574 | 18 |
| 9 | 8 | 17 | 144 | 145 | (17, 144, 145) | 306 | 1224 | 8 |
| 12 | 1 | 143 | 24 | 145 | (24, 143, 145) | 312 | 1716 | 11 |
| 11 | 4 | 105 | 88 | 137 | (88, 105, 137) | 330 | 4620 | 28 |
| 10 | 7 | 51 | 140 | 149 | (51, 140, 149) | 340 | 3570 | 21 |
| 11 | 6 | 85 | 132 | 157 | (85, 132, 157) | 374 | 5610 | 30 |
| 10 | 9 | 19 | 180 | 181 | (19, 180, 181) | 380 | 1710 | 9 |
| 13 | 2 | 165 | 52 | 173 | (52, 165, 173) | 390 | 4290 | 22 |
| 12 | 5 | 119 | 120 | 169 | (119, 120, 169) | 408 | 7140 | 35 |
| 11 | 8 | 57 | 176 | 185 | (57, 176, 185) | 418 | 5016 | 24 |
| 14 | 1 | 195 | 28 | 197 | (28, 195, 197) | 420 | 2730 | 13 |
| 13 | 4 | 153 | 104 | 185 | (104, 153, 185) | 442 | 7956 | 36 |
| 12 | 7 | 95 | 168 | 193 | (95, 168, 193) | 456 | 7980 | 35 |
| 11 | 10 | 21 | 220 | 221 | (21, 220, 221) | 462 | 2310 | 10 |
| 14 | 3 | 187 | 84 | 205 | (84, 187, 205) | 476 | 7854 | 33 |
| 13 | 6 | 133 | 156 | 205 | (133, 156, 205) | 494 | 10374 | 42 |
| 15 | 2 | 224 | 60 | 229 | (60, 221, 229) | 510 | 6630 | 26 |
| 14 | 5 | 171 | 140 | 221 | (140, 171, 221) | 532 | 11970 | 45 |
| 16 | 1 | 255 | 32 | 257 | (32, 255, 257) | 544 | 4080 | 15 |
| 13 | 8 | 105 | 208 | 233 | (105, 208, 233) | 546 | 10920 | 40 |
| 12 | 11 | 23 | 264 | 265 | (23, 264, 265) | 552 | 3036 | 11 |
| 15 | 4 | 209 | 120 | 241 | (120, 209, 241) | 570 | 12540 | 44 |
| 13 | 10 | 69 | 260 | 269 | (69, 260, 269) | 598 | 8970 | 30 |
| 16 | 3 | 247 | 96 | 265 | (96, 247, 265) | 608 | 11856 | 39 |
| 14 | 9 | 115 | 252 | 277 | (115, 252, 277) | 644 | 14490 | 45 |

| $m$ | $n$ | $a = m^2 - n^2$ | $b = 2mn$ | $c = m^2 + n^2$ | Pythagorean Triple | Perimeter | Area | In-radius |
|---|---|---|---|---|---|---|---|---|
| 17 | 2 | 285 | 68 | 293 | (68, 285, 293) | 646 | 9690 | 30 |
| 13 | 12 | 25 | 312 | 313 | (25, 312, 313) | 650 | 3900 | 12 |
| 16 | 5 | 231 | 160 | 281 | (160, 231, 281) | 672 | 18480 | 55 |
| 18 | 1 | 323 | 36 | 325 | (36, 323, 325) | 684 | 5814 | 17 |
| 15 | 8 | 161 | 240 | 289 | (161, 240, 289) | 690 | 19320 | 56 |
| 14 | 11 | 75 | 308 | 317 | (75, 308, 317) | 700 | 11550 | 33 |
| 17 | 4 | 273 | 136 | 305 | (136, 273, 305) | 714 | 714 | 52 |
| 16 | 7 | 207 | 224 | 305 | (207, 224, 305) | 736 | 736 | 63 |
| 14 | 13 | 27 | 364 | 365 | (27,364,365) | 756 | 4914 | 13 |
| 17 | 6 | 253 | 204 | 325 | (204,253,325) | 782 | 25806 | 66 |
| 19 | 2 | 357 | 76 | 365 | (76,357,365) | 798 | 13566 | 34 |
| 16 | 9 | 175 | 288 | 337 | (175,288,337) | 800 | 25200 | 63 |
| 18 | 5 | 299 | 180 | 349 | (180,299,349) | 828 | 26910 | 65 |
| 20 | 1 | 399 | 40 | 401 | (40,399,401) | 840 | 7980 | 19 |
| 17 | 8 | 225 | 272 | 353 | (225,272,353) | 850 | 30600 | 72 |
| 16 | 11 | 135 | 352 | 377 | (135,352,377) | 864 | 23760 | 55 |
| 15 | 14 | 29 | 420 | 421 | (29,420,421) | 870 | 6090 | 14 |
| 19 | 4 | 345 | 152 | 377 | (152,345,377) | 874 | 26220 | 60 |
| 18 | 7 | 275 | 252 | 373 | (252,275,373) | 900 | 34650 | 77 |
| 17 | 10 | 189 | 340 | 389 | (189,340,389) | 918 | 32130 | 70 |
| 20 | 3 | 391 | 120 | 409 | (120,391,409) | 920 | 23460 | 51 |
| 16 | 13 | 87 | 416 | 425 | (87,416,425) | 928 | 18096 | 39 |
| 19 | 6 | 325 | 228 | 397 | (228,325,397) | 950 | 37050 | 78 |
| 21 | 2 | 437 | 84 | 445 | (84,437,445) | 966 | 18354 | 38 |
| 17 | 12 | 145 | 408 | 433 | (145,408,433) | 986 | 29580 | 60 |
| 16 | 15 | 31 | 480 | 481 | (31,480,481) | 992 | 7440 | 15 |
| 22 | 1 | 483 | 44 | 485 | (44,483,485) | 1012 | 10626 | 21 |
| 19 | 8 | 297 | 304 | 425 | (297,304,425) | 1026 | 45144 | 88 |
| 18 | 11 | 203 | 396 | 445 | (203,396,445) | 1044 | 40194 | 77 |
| 21 | 4 | 425 | 168 | 457 | (168,425,457) | 1050 | 35700 | 68 |
| 17 | 14 | 93 | 476 | 485 | (93,476,485) | 1054 | 22134 | 42 |
| 20 | 7 | 351 | 280 | 449 | (280,351,449) | 1080 | 49140 | 91 |
| 22 | 3 | 475 | 132 | 493 | (132,475,493) | 1100 | 31350 | 57 |
| 19 | 10 | 261 | 380 | 461 | (261,380,461) | 1102 | 4950 | 90 |
| 18 | 13 | 155 | 468 | 493 | (155,468,493) | 1116 | 36270 | 65 |
| 17 | 16 | 33 | 544 | 545 | (33,544,545) | 1122 | 8976 | 16 |
| 23 | 2 | 525 | 92 | 533 | (92,525,533) | 1150 | 24150 | 42 |
| 20 | 9 | 319 | 360 | 481 | (319,360,481) | 1160 | 57420 | 99 |
| 19 | 12 | 217 | 456 | 505 | (217,456,505) | 1178 | 49476 | 84 |
| 22 | 5 | 459 | 220 | 509 | (220,459,509) | 1188 | 50490 | 85 |
| 24 | 1 | 575 | 48 | 577 | (48,575,577) | 1200 | 13800 | 23 |
| 21 | 8 | 377 | 336 | 505 | (336,377,505) | 1218 | 63336 | 104 |
| 20 | 11 | 279 | 440 | 521 | (279,440,521) | 1240 | 61380 | 99 |
| 23 | 4 | 513 | 184 | 545 | (184,513,545) | 1242 | 47196 | 76 |
| 19 | 14 | 165 | 532 | 557 | (165,532,557) | 1254 | 10710 | 70 |
| 18 | 17 | 35 | 612 | 613 | (35,612,613) | 1260 | 66990 | 17 |
| 22 | 7 | 435 | 308 | 533 | (308,435,533) | 1276 | 71610 | 105 |
| 21 | 10 | 341 | 420 | 541 | (341,420,541) | 1302 | 60060 | 110 |
| 20 | 13 | 231 | 520 | 569 | (231,520,569) | 1320 | 31920 | 91 |
| 19 | 16 | 105 | 608 | 617 | (105,608,617) | 1330 | 68034 | 48 |
| 23 | 6 | 493 | 276 | 565 | (276,493,565) | 1334 | 31050 | 102 |
| 25 | 2 | 621 | 100 | 629 | (100,621,629) | 1350 | 79794 | 46 |
| 22 | 9 | 403 | 396 | 565 | (396,403,565) | 1364 | 66120 | 117 |
| 24 | 5 | 551 | 240 | 601 | (240,551,601) | 1392 | 17550 | 95 |
| 26 | 1 | 675 | 52 | 677 | (52,675,677) | 1404 | 12654 | 25 |
| 19 | 18 | 37 | 684 | 685 | (37,684,685) | 1406: | 12654 | 18 |
| 23 | 8 | 465 | 368 | 593 | (368,465,593) | 1426 | 85560 | 120 |
| 25 | 4 | 609 | 200 | 641 | (200,609,641) | 1450 | 60900 | 84 |
| 20 | 17 | 111 | 680 | 689 | (111,680,689) | 1480 | 37740 | 51 |
| 24 | 7 | 527 | 336 | 625 | (336,527,625) | 1488 | 88536 | 119 |
| 26 | 3 | 667 | 156 | 685 | (156,667,685) | 1508 | 52026 | 69 |
| 23 | 10 | 429 | 460 | 629 | (429,460,629) | 1518 | 98670 | 130 |
| 22 | 13 | 315 | 572 | 653 | (315,572,653) | 1540 | 90090 | 117 |
| 25 | 6 | 589 | 300 | 661 | (300,589,661) | 1550 | 88350 | 114 |
| 21 | 16 | 185 | 672 | 697 | (185,672,697) | 1554 | 62160 | 80 |
| 20 | 19 | 39 | 760 | 761 | (39,760,761) | 1560 | 14820 | 19 |
| 27 | 2 | 725 | 108 | 733 | (108,725,733) | 1566 | 39150 | 50 |
| 23 | 12 | 385 | 552 | 673 | (385,552,673) | 1610 | 106260 | 132 |
| 26 | 5 | 651 | 260 | 701 | (260,651,701) | 1612 | 84630 | 105 |
| 28 | 1 | 783 | 56 | 785 | (56,783,785) | 1624 | 21924 | 27 |
| 22 | 15 | 259 | 660 | 709 | (259,660,709) | 1628 | 85470 | 105 |
| 25 | 8 | 561 | 400 | 689 | (400,561,689) | 1650 | 112200 | 136 |
| 27 | 4 | 713 | 216 | 745 | (216,713,745) | 1674 | 77004 | 92 |
| 24 | 11 | 455 | 528 | 697 | (455,528,697) | 1680 | 120120 | 143 |
| 23 | 14 | 333 | 644 | 725 | (333,644,725) | 1702 | 107226 | 126 |

| $m$ | $n$ | $a = m^2 - n^2$ | $b = 2mn$ | $c = m^2 + n^2$ | Pythagorean Triple | Peri-meter | Area | In-radius |
|---|---|---|---|---|---|---|---|---|
| 22 | 17 | 195 | 748 | 773 | (195,748,773) | 1716 | 72930 | 85 |
| 26 | 7 | 627 | 364 | 725 | (364,627,725) | 1716 | 114114 | 133 |
| 21 | 20 | 41 | 840 | 841 | (41,840,841) | 1722 | 17220 | 20 |
| 28 | 3 | 775 | 168 | 793 | (168,775,793) | 1736 | 65100 | 75 |
| 24 | 13 | 407 | 624 | 745 | (407,624,745) | 1776 | 126984 | 143 |
| 23 | 16 | 273 | 736 | 785 | (273,736,785) | 1794 | 100464 | 112 |
| 29 | 2 | 837 | 116 | 845 | (116,837,845) | 1798 | 48546 | 54 |
| 22 | 19 | 123 | 836 | 845 | (123,836,845) | 1804 | 51414 | 57 |
| 26 | 9 | 595 | 468 | 757 | (468,595,757) | 1820 | 139230 | 153 |
| 28 | 5 | 759 | 280 | 809 | (280,759,809) | 1848 | 106260 | 115 |
| 25 | 12 | 481 | 600 | 769 | (481,600,769) | 1850 | 144300 | 156 |
| 30 | 1 | 899 | 60 | 901 | (60,899,901) | 1860 | 26970 | 29 |
| 23 | 18 | 205 | 828 | 853 | (205,828,853) | 1886 | 84870 | 90 |
| 27 | 8 | 665 | 432 | 793 | (432,665,793) | 1890 | 143640 | 152 |
| 22 | 21 | 43 | 924 | 925 | (43,924,925) | 1892 | 19866 | 21 |
| 29 | 4 | 825 | 232 | 857 | (232,825,857) | 1914 | 95700 | 100 |
| 26 | 11 | 555 | 572 | 797 | (555,572,797) | 1924 | 158730 | 165 |
| 25 | 14 | 429 | 700 | 821 | (429,700,821) | 1950 | 150150 | 154 |
| 24 | 17 | 287 | 816 | 865 | (287,816,865) | 1968 | 117096 | 119 |
| 23 | 20 | 129 | 920 | 929 | (129,920,929) | 1978 | 59340 | 60 |
| 27 | 10 | 629 | 540 | 829 | (540,629,829) | 1998 | 169830 | 170 |
| 29 | 6 | 805 | 348 | 877 | (348,805,877) | 2030 | 140070 | 138 |
| 31 | 2 | 957 | 124 | 965 | (124,957,965) | 2046 | 59334 | 58 |
| 25 | 16 | 369 | 800 | 881 | (369,800,881) | 2050 | 147600 | 144 |
| 24 | 19 | 215 | 912 | 937 | (215,912,937) | 2064 | 98040 | 95 |
| 23 | 22 | 45 | 1012 | 1013 | (45,1012,1013) | 2070 | 22770 | 22 |
| 28 | 9 | 703 | 504 | 865 | (504,703,865) | 2072 | 177156 | 171 |
| 32 | 1 | 1023 | 64 | 1025 | (64,1023,1025) | 2112 | 32736 | 31 |
| 26 | 15 | 451 | 780 | 901 | (451,780,901) | 2132 | 175890 | 165 |
| 29 | 8 | 777 | 464 | 905 | (464,777,905) | 2146 | 180264 | 168 |
| 25 | 18 | 301 | 900 | 949 | (301,900,949) | 2150 | 135450 | 126 |
| 31 | 4 | 945 | 248 | 977 | (248,945,977) | 2170 | 117180 | 108 |
| 28 | 11 | 663 | 616 | 905 | (616,663,905) | 2184 | 204204 | 187 |
| 27 | 14 | 533 | 756 | 925 | (533,756,925) | 2214 | 201474 | 182 |
| 30 | 7 | 851 | 420 | 949 | (420,851,949) | 2220 | 178710 | 161 |
| 26 | 17 | 387 | 884 | 965 | (387,884,965) | 2236 | 171054 | 153 |
| 32 | 3 | 1015 | 192 | 1033 | (192,1015,1033) | 2240 | 97440 | 87 |
| 24 | 23 | 47 | 1104 | 1105 | (47,1104,1105) | 2256 | 25944 | 23 |
| 29 | 10 | 741 | 580 | 941 | (580,741,941) | 2262 | 214890 | 190 |
| 31 | 6 | 925 | 372 | 997 | (372,925,997) | 2294 | 172050 | 150 |
| 28 | 13 | 615 | 728 | 953 | (615,728,953) | 2296 | 223860 | 195 |
| 33 | 2 | 1085 | 132 | 1093 | (132,1085,1093) | 2310 | 71610 | 62 |
| 27 | 16 | 473 | 864 | 985 | (473,864,985) | 2322 | 204336 | 176 |
| 26 | 19 | 315 | 988 | 1037 | (315,988,1037) | 2340 | 155610 | 133 |
| 25 | 22 | 141 | 1100 | 1109 | (141,1100,1109) | 2350 | 204336 | 66 |
| 32 | 5 | 999 | 320 | 1049 | (320,999,1049) | 2368 | 155610 | 135 |
| 26 | 12 | 697 | 696 | 985 | (696,697,985) | 2378 | 77550 | 204 |
| 34 | 1 | 1155 | 68 | 1157 | (68,1155,1157) | 2380 | 39270 | 33 |
| 28 | 15 | 559 | 840 | 1009 | (559, 840, 1009) | 2408 | 234780 | 195 |
| 33 | 4 | 897 | 496 | 1025 | (496, 897, 1025) | 2418 | 222456 | 184 |
| 26 | 21 | 1073 | 264 | 1105 | (264, 1073, 1105) | 2442 | 141636 | 116 |
| 25 | 24 | 49 | 1200 | 1201 | (49,1200,1201) | 2450 | 29400 | 24 |
| 30 | 11 | 779 | 660 | 1021 | (660,779,1021) | 2460 | 257070 | 209 |
| 29 | 14 | 645 | 812 | 1037 | (645,812,1037) | 2494 | 261870 | 210 |
| 32 | 7 | 975 | 448 | 1073 | (448,975,1073) | 2496 | 218400 | 175 |
| 34 | 3 | 1147 | 204 | 1165 | (204,1147,1165) | 2516 | 116994 | 93 |
| 28 | 17 | 495 | 952 | 1073 | (495,952,1073) | 2520 | 235620 | 187 |
| 27 | 20 | 329 | 1080 | 1129 | (329,1080,1129) | 2538 | 177660 | 140 |
| 31 | 10 | 861 | 620 | 1061 | (620,861,1061) | 2542 | 266910 | 210 |
| 26 | 23 | 147 | 1196 | 1205 | (147,1196,1205) | 2548 | 87906 | 69 |
| 30 | 13 | 731 | 780 | 1069 | (731,780,1069) | 2580 | 285090 | 221 |
| 35 | 2 | 1221 | 140 | 1229 | (140,1221,1229) | 2590 | 85470 | 66 |
| 35 | 16 | 585 | 928 | 1097 | (585,928,1097) | 2610 | 271440 | 208 |
| 29 | 9 | 943 | 576 | 1105 | (576,943,1105) | 2624 | 271584 | 207 |
| 32 | 19 | 423 | 1064 | 1145 | (423,1064,1145) | 2632 | 225036 | 171 |
| 28 | 22 | 245 | 1188 | 1213 | (245,1188,1213) | 2646 | 145530 | 110 |
| 27 | 25 | 51 | 1300 | 1301 | (51,1300,1301) | 2652 | 33150 | 25 |
| 26 | 5 | 1131 | 340 | 1181 | (340,1131,1181) | 2652 | 192270 | 145 |
| 34 | 1 | 1295 | 72 | 1297 | (72,1295,1297) | 2664 | 46620 | 35 |
| 36 | 12 | 817 | 744 | 1105 | (744,817,1105) | 2666 | 303924 | 228 |
| 31 | 8 | 1025 | 528 | 1153 | (528,1025,1153) | 2706 | 270600 | 200 |
| 33 | 18 | 517 | 1044 | 1165 | (517,1044,1165) | 2726 | 269874 | 198 |
| 29 | 4 | 1209 | 280 | 1241 | (280,1209,1241) | 2730 | 169260 | 124 |
| 35 | 11 | 903 | 704 | 1145 | (704,903,1145) | 2752 | 317856 | 231 |
| 32 | 7 | 1107 | 476 | 1205 | (476,1107,1205) | 2788 | 263466 | 189 |

| $m$ | $n$ | $a = m^2 - n^2$ | $b = 2mn$ | $c = m^2 + n^2$ | Pythagorean Triple | Perimeter | Area | In-radius |
|---|---|---|---|---|---|---|---|---|
| 34 | 14 | 765 | 868 | 1157 | (765,868,1157) | 2790 | 332010 | 238 |
| 30 | 17 | 611 | 1020 | 1189 | (611,1020,1189) | 2820 | 311610 | 221 |
| 33 | 10 | 989 | 660 | 1189 | (660,989,1189) | 2838 | 326370 | 230 |
| 29 | 20 | 441 | 1160 | 1241 | (441,1160,1241) | 2842 | 255780 | 180 |
| 28 | 23 | 255 | 1288 | 1313 | (255,1288,1313) | 2856 | 164220 | 115 |
| 27 | 26 | 53 | 1404 | 1405 | (53,1404,1405) | 2862 | 37206 | 26 |
| 35 | 6 | 1189 | 420 | 1261 | (420,1189,1261) | 2870 | 249690 | 174 |
| 32 | 13 | 855 | 832 | 1193 | (832,855,1193) | 2880 | 355680 | 247 |
| 37 | 2 | 1365 | 148 | 1373 | (148,1365,1373) | 2886 | 101010 | 70 |
| 31 | 16 | 705 | 332 | 1217 | (705,992,1217) | 2914 | 349680 | 240 |
| 34 | 9 | 1075 | 612 | 1237 | (612,1075,1237) | 2924 | 328950 | 225 |
| 30 | 19 | 539 | 1140 | 1261 | (539,1140,1261) | 2940 | 307230 | 209 |
| 36 | 5 | 1271 | 360 | 1321 | (360,1271,1321) | 2952 | 228780 | 155 |
| 29 | 22 | 1276 | 1276 | 1325 | (357,1276,1325) | 2958 | 227766 | 154 |
| 38 | 1 | 1443 | 76 | 1445 | (76,1443,1445) | 2964 | 54834 | 37 |
| 28 | 25 | 159 | 1400 | 1409 | (159,1400,140) | 2968 | 111300 | 75 |
| 32 | 15 | 799 | 960 | 1249 | (799,960,1249) | 3008 | 383520 | 255 |
| 35 | 8 | 1161 | 560 | 1289 | (560,1161,1289) | 3010 | 325080 | 216 |
| 37 | 4 | 1353 | 296 | 1385 | (296,1353,1385) | 3034 | 200244 | 132 |
| 31 | 18 | 637 | 1116 | 1285 | (637,1116,1285) | 3038 | 355446 | 234 |
| 34 | 11 | 1035 | 748 | 1277 | (748,1035,1277) | 3060 | 387090 | 253 |
| 29 | 24 | 265 | 7392 | 1417 | (265,1392 1417) | 3074 | 184440 | 120 |
| 28 | 27 | 55 | 1512 | 1513 | (55.1512,1513) | 3080 | 41580 | 27 |
| 36 | 7 | 1247 | 504 | 1345 | (504,1247,1345) | 3096 | 314244 | 203 |
| 33 | 14 | 893 | 924 | 1285 | (893,924,1285) | 3102 | 412566 | 266 |
| 38 | 3 | 1435 | 228 | 1453 | (228,1435,1453) | 3116 | 163590 | 105 |
| 32 | 17 | 735 | 1088 | 1313 | (735,1088,1313) | 3136 | 399840 | 255 |
| 31 | 20 | 561 | 1240 | 1361 | (561,1240,1361) | 3162 | 347820 | 220 |
| 30 | 23 | 371 | 1380 | 1429 | (371,1380,1429) | 3180 | 255990 | 161 |
| 37 | 6 | 1333 | 444 | 1405 | (444,1333,1405) | 3182 | 295926 | 186 |
| 29 | 26 | 165 | 1508 | 1517 | (165,1508.1517) | 3190 | 124410 | 29 |
| 34 | 13 | 987 | 884 | 1325 | (884,987,1325) | 3196 | 436254 | 273 |
| 39 | 2 | 1517 | 156 | 1525 | (156.1517.1525) | 3198 | 118326 | 74 |
| 33 | 16 | 833 | 1056 | 1345 | (833,1056,1345) | 3234 | 439824 | 272 |
| 32 | 19 | 663 | 1216 | 1385 | (663,1216,1385) | 3264 | 403104 | 247 |
| 38 | 5 | 1419 | 1419 | 1469 | (380,1419,1469) | 3268 | 269610 | 165 |
| 40 | 1 | 1599 | 80 | 1601 | (80,1599,1601) | 3280 | 63960 | 39 |
| 31 | 22 | 1364 | 1364 | 1445 | (477,1364.1445) | 3286 | 325314 | 198 |
| 35 | 12 | 1081 | 840 | 1369 | (840,1081,1369) | 3290 | 454020 | 276 |
| 29 | 28 | 57 | 1624 | 1625 | (57,1624,1625) | 3306 | 46284 | 28 |
| 37 | 8 | 1305 | 592 | 1433 | (592,1305,1433) | 3330 | 386280 | 232 |
| 34 | 15 | 931 | 1020 | 1381 | (931,1020.1381) | 3332 | 474810 | 285 |
| 39 | 4 | 1505 | 312 | 1537 | (312,1505.1537) | 3354 | 234780 | 140 |
| 36 | 11 | 1175 | 792 | 1417 | (792,1175.1417) | 3384 | 465300 | 275 |
| 32 | 21 | 583 | 1344 | 1465 | (583.1344.1465) | 3392 | 391776 | 231 |
| 31 | 24 | 385 | 1488 | 1537 | (385,1488,1537) | 3410 | 286440 | 168 |
| 38 | 7 | 1395 | 532 | 1493 | (532.1395.1493) | 3420 | 371070 | 217 |
| 40 | 3 | 1591 | 240 | 1609 | (240,1591,1609) | 3440 | 190920 | 111 |
| 37 | 10 | 1269 | 1269 | 1469 | (740,1269,1469) | 3478 | 469530 | 270 |
| 33 | 20 | 689 | 1320 | 1489 | (689,1320,1489) | 3498 | 454740 | 260 |
| 32 | 23 | 495 | 1472 | 1553 | (495.1472.1553) | 3520 | 364320 | 207 |
| 41 | 2 | 1677 | 164 | 1685 | (164.1677.1685) | 3526 | 137514 | 78 |
| 36 | 13 | 1127 | 936 | 1465 | (936,1127,1465) | 3528 | 527436 | 299 |
| 31 | 26 | 285 | 1612 | 1637 | (285.1612,1637) | 3534 | 229710 | 130 |
| 30 | 29 | 59 | 1740 | 1741 | (59.1740,1741) | 3540 | 51330 | 29 |
| 35 | 16 | 969 | 1120 | 1481 | (969.1120,1481) | 3570 | 542640 | 304 |
| 38 | 9 | 1363 | 684 | 1525 | (684.1363,1525) | 3572 | 466146 | 261 |
| 34 | 19 | 1292 | 1292 | 1517 | (795.1292,1517) | 3604 | 513570 | 285 |
| 42 | 1 | 1763 | 84 | 1765 | (84,1763,1765) | 3612 | 74046 | 41 |
| 37 | 12 | 1225 | 888 | 1513 | (888.1225.1513) | 3626 | 543900 | 300 |
| 32 | 25 | 399 | 1600 | 1649 | (399,1600.1649) | 3648 | 319200 | 175 |
| 31 | 28 | 177 | 1736 | 1745 | (177.1736,1745) | 3658 | 153636 | 84 |
| 39 | 8 | 1457 | 624 | 1585 | (624.1457.1585) | 3666 | 454584 | 248 |
| 41 | 4 | 1665 | 328 | 1697 | (328.1665.1697) | 3690 | 273060 | 148 |
| 35 | 18 | 901 | 1260 | 1549 | (901.1260.1549) | 3710 | 567630 | 306 |
| 38 | 11 | 836 | 836 | 1565 | (836.1323.1565) | 3724 | 553014 | 297 |
| 34 | 21 | 715 | 1428 | 1597 | (715.1428.1597) | 3740 | 510510 | 273 |
| 40 | 7 | 1551 | 560 | 1649 | (560.1551.1649) | 3760 | 434280 | 231 |
| 37 | 14 | 1173 | 1036 | 1565 | (1036,1173.1565) | 3774 | 607614 | 322 |
| 32 | 27 | 295 | 1728 | 1753 | (295.1728.1753) | 3776 | 254880 | 135 |
| 31 | 30 | 61 | 1860 | 1861 | (61.1860.1861) | 3782 | 56730 | 30 |
| 36 | 17 | 1007 | 1224 | 1585 | (1007.1224.1585) | 3816 | 616284 | 323 |
| 39 | 10 | 1421 | 780 | 1621 | (780.1421,1621) | 3822 | 554190 | 290 |
| 41 | 6 | 1645 | 492 | 1717 | (492.1645.1717) | 3854 | 404670 | 210 |
| 43 | 2 | 1845 | 172 | 1853 | (172.1845.1853) | 3870 | 158670 | 82 |

| $m$ | $n$ | $a = m^2 - n^2$ | $b = 2mn$ | $c = m^2 + n^2$ | Pythagorean Triple | Perimeter | Area | In-radius |
|---|---|---|---|---|---|---|---|---|
| 38 | 13 | 1275 | 988 | 1613 | (988,1275,1613) | 3876 | 629850 | 325 |
| 34 | 23 | 627 | 1564 | 1685 | (627,1564,1685) | 3876 | 490314 | 253 |
| 33 | 26 | 413 | 1716 | 1765 | (413,1716,1765) | 3894 | 354354 | 182 |
| 32 | 29 | 183 | 1856 | 1865 | (183,1856,1865) | 3904 | 169824 | 87 |
| 40 | 9 | 1519 | 720 | 1681 | (720,1519,1681) | 3920 | 546840 | 279 |
| 37 | 16 | 1113 | 1184 | 1625 | (1113, 1184, 1625) | 3922 | 658896 | 336 |
| 42 | 5 | 1739 | 420 | 1789 | (420,1739,1789) | 3948 | 365190 | 185 |
| 36 | 19 | 935 | 1368 | 1657 | (935,1368,1657) | 3960 | 639540 | 323 |
| 44 | 1 | 1935 | 88 | 1937 | (88,1935,1937) | 3960 | 85140 | 43 |
| 35 | 22 | 741 | 1540 | 1709 | (741,1540,1709) | 3990 | 570570 | 286 |
| 34 | 25 | 531 | 1700 | 1781 | (531,1700,1781) | 4012 | 451350 | 225 |
| 41 | 8 | 1617 | 656 | 1745 | (656,1617,1745) | 4018 | 530376 | 264 |
| 33 | 28 | 305 | 1848 | 1873 | (305,1848,1873) | 4026 | 281820 | 140 |
| 38 | 15 | 1213 | 1140 | 1669 | (1140,1219,1669) | 4028 | 694830 | 345 |
| 32 | 31 | 63 | 1984 | 1985 | (63,1984,1985) | 4032 | 62496 | 31 |
| 43 | 4 | 1833 | 344 | 1865 | (344,1833,1865) | 4042 | 315276 | 156 |
| 37 | 18 | 1045 | 1332 | 1693 | (1045,1332,1693) | 4070 | 695970 | 342 |
| 40 | 11 | 1479 | 880 | 1721 | (880.1479.1721) | 4080 | 650760 | 319 |
| 35 | 24 | 649 | 1680 | 1801 | (649,1680,1801) | 4130 | 545160 | 264 |
| 39 | 14 | 1325 | 1092 | 1717 | (1092,1325,1717) | 4134 | 723450 | 350 |
| 44 | 3 | 1927 | 264 | 1945 | (264,1927,1945) | 4136 | 254364 | 123 |
| 34 | 27 | 427 | 1836 | 1885 | (427.1836.1885) | 4148 | 391986 | 189 |
| 38 | 17 | 1155 | 1292 | 1733 | (1155,1292,1733) | 4180 | 746130 | 357 |
| 41 | 10 | 1581 | 820 | 1781 | (820,1581,1781) | 4182 | 648210 | 310 |
| 43 | 6 | 1813 | 516 | 1885 | (516,1813,1885) | 4214 | 467754 | 222 |
| 37 | 20 | 969 | 1480 | 1769 | (969,1480,1769) | 4218 | 717060 | 340 |
| 45 | 2 | 2021 | 180 | 2029 | (180.2021,2029) | 4230 | 181890 | 86 |
| 40 | 13 | 1431 | 1040 | 1769 | (1040,1431,1769) | 4240 | 744120 | 351 |
| 36 | 23 | 767 | 1656 | 1825 | (767,1656,1825) | 4248 | 635076 | 299 |
| 35 | 26 | 549 | 1820 | 1901 | (549,1820,1901) | 4270 | 499590 | 234 |
| 34 | 29 | 315 | 1972 | 1997 | (315,1972,1997) | 4284 | 310590 | 145 |
| 33 | 32 | 65 | 2112 | 2113 | (65,2112,2113) | 4290 | 68640 | 32 |
| 39 | 16 | 1265 | 1248 | 1777 | (1248.1265.1777) | 4290 | 789360 | 368 |
| 44 | 5 | 1911 | 440 | 1961 | (440,1911,1961) | 4312 | 420420 | 195 |
| 46 | 1 | 2115 | 92 | 2117 | (92,2115,2117) | 4324 | 97290 | 45 |
| 41 | 12 | 1537 | 984 | 1825 | (984,1537,1825) | 4346 | 756204 | 348 |
| 37 | 22 | 885 | 1628 | 1853 | (885.1628,1853) | 4366 | 720390 | 330 |
| 43 | 8 | 1785 | 688 | 1913 | (688.1785.1913) | 4386 | 614040 | 280 |
| 36 | 25 | 671 | 1800 | 1921 | (671.1800.1921) | 4392 | 603900 | 275 |
| 45 | 4 | 2009 | 360 | 2041 | (360.2009,2041) | 4410 | 361620 | 164 |
| 34 | 31 | 195 | 2108 | 2117 | (195,2108,2117) | 4420 | 205530 | 93 |
| 42 | 11 | 1643 | 924 | 1885 | (924. 1643. 1885) | 4452 | 759066 | 341 |
| 38 | 21 | 1003 | 1596 | 1885 | (1003. 1596. 1885) | 4484 | 800394 | 357 |
| 44 | 7 | 1887 | 616 | 1985 | (616.1887,1985) | 4488 | 581196 | 259 |
| 46 | 3 | 2107 | 276 | 2125 | (276.2107.2125) | 4508 | 290766 | 129 |
| 41 | 14 | 1485 | 1148 | 1877 | (1148,1485,1877) | 4510 | 852390 | 378 |
| 37 | 24 | 793 | 1776 | 1945 | (793 1776.1945) | 4514 | 704184 | 312 |
| 34 | 33 | 67 | 2244 | 2245 | (67,2244.2245) | 4556 | 75174 | 33 |
| 43 | 10 | 1749 | 860 | 1949 | (860,1749,1949) | 4558 | 752070 | 330 |
| 40 | 17 | 1311 | 1360 | 1889 | (1311,1360.1889) | 4560 | 891480 | 391 |
| 39 | 20 | 1121 | 1560 | 1921 | (1121.1560.1921) | 4602 | 874380 | 380 |
| 47 | 2 | 2205 | 188 | 2213 | (188.2205.2213) | 4606 | 207270 | 90 |
| 42 | 13 | 1595 | 1092 | 1933 | (1092.1595,1933) | 4620 | 870870 | 377 |
| 38 | 23 | 915 | 1748 | 1973 | (915.1748,1973) | 4636 | 799710 | 345 |
| 37 | 26 | 693 | 1924 | 2045 | (693.1924.2045) | 4662 | 666666 | 286 |
| 44 | 9 | 1855 | 792 | 2017 | (792.1855,2017) | 4664 | 734580 | 315 |
| 41 | 16 | 1425 | 1312 | 1937 | (1312.1425.1937) | 4674 | 934800 | 400 |
| 36 | 29 | 455 | 2088 | 2137 | (455,2088.2137) | 4680 | 475020 | 203 |
| 35 | 32 | 201 | 2240 | 2249 | (201.2240.2249) | 4690 | 225120 | 96 |
| 46 | 5 | 2091 | 460 | 2141 | (460.2091.2141) | 4692 | 480930 | 205 |
| 48 | 1 | 2303 | 96 | 2305 | (96.2303.2305) | 4704 | 110544 | 47 |
| 40 | 19 | 1239 | 1520 | 1961 | (1239.1520.1961) | 4720 | 941640 | 399 |
| 43 | 12 | 1705 | 1032 | 1993 | (1032.1705.1993) | 4730 | 879780 | 372 |
| 39 | 22 | 1037 | 1716 | 2005 | (1037.1716.2005) | 4758 | 889746 | 374 |
| 45 | 8 | 1961 | 720 | 2089 | (720.1961,2089) | 4770 | 705960 | 296 |
| 38 | 25 | 819 | 1900 | 2069 | (819.1900.2069) | 4788 | 778050 | 325 |
| 47 | 4 | 2193 | 376 | 2225 | (376.2193.2225) | 4794 | 412284 | 172 |
| 37 | 28 | 585 | 2072 | 2153 | (585.2072.2153) | 4810 | 606060 | 252 |
| 36 | 31 | 335 | 2232 | 2257 | (335.2232.2257) | 4824 | 373860 | 155 |
| 35 | 34 | 69 | 2380 | 2381 | (69.2380,2381) | 4830 | 82110 | 34 |
| 41 | 18 | 1357 | 1476 | 2005 | (1357.1476.2005) | 4838 | 1001466 | 414 |
| 46 | 7 | 2067 | 644 | 2165 | (644.2067.2165) | 4876 | 665574 | 273 |
| 40 | 21 | 1159 | 1680 | 2041 | (1159.1680.2041) | 4880 | 973560 | 399 |
| 43 | 14 | 1653 | 1204 | 2045 | (1204.1653.2045) | 4902 | 995106 | 406 |
| 38 | 27 | 715 | 2052 | 2173 | (715.2052.2173) | 4940 | 733590 | 297 |

| $m$ | $n$ | $a = m^2 - n^2$ | $b = 2mn$ | $c = m^2 + n^2$ | **Pythagorean Triple** | **Peri-meter** | **Area** | **In-radius** |
|---|---|---|---|---|---|---|---|---|
| 42 | 17 | 1475 | 1428 | 2053 | (1428,1475,2053) | 4956 | 1053150 | 425 |
| 37 | 30 | 469 | 2220 | 2269 | (469,2220,2269) | 4958 | 520590 | 210 |
| 47 | 6 | 2173 | 564 | 2245 | (564,2173,2245) | 4982 | 612786 | 246 |
| 49 | 2 | 2397 | 196 | 2405 | (196,2397,2405) | 4998 | 234906 | 94 |

# Appendix D

# List of Pythagorean Triples—Primitive and Nonprimitive

| $m$ | $n$ | $a = m^2 - n^2$ | $b = 2mn$ | $c = m^2 + n^2$ | *Primitive or multiple* |
|---|---|---|---|---|---|
| 2 | 1 | 3 | 4 | 5 | Primitive |
| 3 | 1 | 8 | 6 | 10 | 2x(4, 3, 5) |
| 3 | 2 | 5 | 12 | 13 | Primitive |
| 4 | 1 | 15 | 8 | 17 | Primitive |
| 4 | 2 | 12 | 16 | 20 | 4x(3, 4, 5) |
| 4 | 3 | 7 | 24 | 25 | Primitive |
| 5 | 1 | 24 | 10 | 26 | 2x(12, 5, 13) |
| 5 | 2 | 21 | 20 | 29 | Primitive |
| 5 | 3 | 16 | 30 | 34 | 2x(8, 15, 17) |
| 5 | 4 | 9 | 40 | 41 | Primitive |
| 6 | 1 | 35 | 12 | 37 | Primitive |
| 6 | 2 | 32 | 24 | 40 | 8x(4. 3. 5) |
| 6 | 3 | 27 | 36 | 45 | 9x(3, 4, 5) |
| 6 | 4 | 20 | 48 | 52 | 4x(5, 12, 13) |
| 6 | 5 | 11 | 60 | 61 | Primitive |
| 7 | 1 | 48 | 14 | 50 | 2x(24. 7. 25) |
| 7 | 2 | 45 | 28 | 53 | Primitive |
| 7 | 3 | 40 | 42 | 58 | 2x(20, 21. 29) |
| 7 | 4 | 33 | 56 | 65 | Primitive |
| 7 | 5 | 24 | 70 | 74 | 2x(12. 35. 37) |
| 7 | 6 | 13 | 84 | 85 | Primitive |
| 8 | 1 | 63 | 16 | 65 | Primitive |
| 8 | 2 | 60 | 32 | 68 | 4x(15. 8. 17) |
| 8 | 3 | 55 | 48 | 73 | Primitive |
| 8 | 4 | 48 | 64 | 80 | 16x(3, 4. 5) |
| 8 | 5 | 39 | 80 | 89 | Primitive |
| 8 | 6 | 28 | 96 | 100 | 4x(7. 24. 25) |
| 8 | 7 | 15 | 112 | 113 | Primitive |
| 9 | 1 | 80 | 18 | 82 | 2x(40. 9, 41) |
| 9 | 2 | 77 | 36 | 85 | Primitive |
| 9 | 3 | 72 | 54 | 90 | 18x(4. 3. 5) |
| 9 | 4 | 65 | 72 | 97 | Primitive |

| $m$ | $n$ | $a = m^2 - n^2$ | $b = 2mn$ | $c = m^2 + n^2$ | *Primitive or multiple* |
|---|---|---|---|---|---|
| 9 | 5 | 56 | 90 | 106 | 2x(28, 45, 53) |
| 9 | 6 | 45 | 108 | 117 | 9x(5, 12, 13) |
| 9 | 7 | 32 | 126 | 130 | 2x(16, 63, 65) |
| 9 | 8 | 17 | 144 | 145 | Primitive |
| 10 | 1 | 99 | 20 | 101 | Primitive |
| 10 | 2 | 96 | 40 | 104 | 8x(12, 5, 13) |
| 10 | 3 | 91 | 60 | 109 | Primitive |
| 10 | 4 | 84 | 80 | 116 | 4x(21, 20, 29) |
| 10 | 5 | 75 | 100 | 125 | 25x(3, 4, 5) |
| 10 | 6 | 64 | 120 | 136 | 8x(8, 15, 17) |
| 10 | 7 | 51 | 140 | 149 | Primitive |
| 10 | 8 | 36 | 160 | 164 | 4x(9, 40, 41) |
| 10 | 9 | 19 | 180 | 181 | Primitive |
| 11 | 1 | 120 | 22 | 122 | 2x(60, 11, 61) |
| 11 | 2 | 117 | 44 | 125 | Primitive |
| 11 | 3 | 112 | 66 | 130 | 2x(56, 33, 65) |
| 11 | 4 | 105 | 88 | 137 | Primitive |
| 11 | 5 | 96 | 110 | 146 | 2x(48, 55, 73) |
| 11 | 6 | 85 | 132 | 157 | Primitive |
| 11 | 7 | 72 | 154 | 170 | 2x(36, 77, 85) |
| 11 | 8 | 57 | 176 | 185 | Primitive |
| 11 | 9 | 40 | 198 | 202 | 2x(20, 99, 101) |
| 11 | 10 | 21 | 220 | 221 | Primitive |
| 12 | 1 | 143 | 24 | 145 | Primitive |
| 12 | 2 | 140 | 48 | 148 | 4x(35, 12, 37) |
| 12 | 3 | 135 | 72 | 153 | 9x(15, 8, 17) |
| 12 | 4 | 128 | 96 | 160 | 32x(4, 3, 5) |
| 12 | 5 | 119 | 120 | 169 | Primitive |
| 12 | 6 | 108 | 144 | 180 | 36x(3, 4, 5) |
| 12 | 7 | 95 | 168 | 193 | Primitive |
| 12 | 8 | 80 | 192 | 208 | 16x(5, 12, 13) |
| 12 | 9 | 63 | 216 | 225 | 9x(7, 24, 25) |
| 12 | 10 | 44 | 240 | 244 | 4x(11, 60, 61) |
| 12 | 11 | 23 | 264 | 265 | Primitive |
| 13 | 1 | 168 | 26 | 170 | 2x(84, 13, 85) |
| 13 | 2 | 165 | 52 | 173 | Primitive |
| 13 | 3 | 160 | 78 | 178 | 2x(80, 39, 89) |
| 13 | 4 | 153 | 104 | 185 | Primitive |
| 13 | 5 | 144 | 130 | 194 | 2x(72, 65, 97) |
| 13 | 6 | 133 | 156 | 205 | Primitive |
| 13 | 7 | 120 | 182 | 218 | 2x(60, 91, 109) |
| 13 | 8 | 105 | 208 | 233 | Primitive |
| 13 | 9 | 88 | 234 | 250 | 2x(44, 11, 125) |
| 13 | 10 | 69 | 260 | 269 | Primitive |
| 13 | 11 | 48 | 286 | 290 | 2x(24, 143, 145) |
| 13 | 12 | 25 | 312 | 313 | Primitive |
| 14 | 1 | 195 | 28 | 197 | Primitive |
| 14 | 2 | 192 | 56 | 200 | 8x(24, 7, 25) |
| 14 | 3 | 187 | 84 | 205 | Primitive |
| 14 | 4 | 180 | 112 | 212 | 4x(45, 28, 53) |
| 14 | 5 | 171 | 140 | 221 | Primitive |
| 14 | 6 | 160 | 168 | 232 | 8x(20, 21, 29) |
| 14 | 7 | 147 | 196 | 245 | 49x(3, 4, 5) |
| 14 | 8 | 132 | 224 | 260 | 4x(33, 56, 65) |
| 14 | 9 | 115 | 252 | 277 | Primitive |
| 14 | 10 | 96 | 280 | 296 | 8x(12, 35, 37) |

| $m$ | $n$ | $a = m^2 - n^2$ | $b = 2mn$ | $c = m^2 + n^2$ | *Primitive or multiple* |
|---|---|---|---|---|---|
| 14 | 11 | 75 | 308 | 317 | Primitive |
| 14 | 12 | 52 | 336 | 340 | 4x(13, 84, 85) |
| 14 | 13 | 27 | 364 | 365 | Primitive |
| 15 | 1 | 224 | 30 | 226 | 2x(112, 15, 113) |
| 15 | 2 | 221 | 60 | 229 | Primitive |
| 15 | 3 | 216 | 90 | 234 | 18x(12. 5. 13) |
| 15 | 4 | 209 | 120 | 241 | Primitive |
| 15 | 5 | 200 | 150 | 250 | 50x(4, 3, 5) |
| 15 | 6 | 189 | 180 | 261 | 9x(21, 20, 29) |
| 15 | 7 | 176 | 210 | 274 | 2x(88, 105, 137) |
| 15 | 8 | 161 | 240 | 289 | Primitive |
| 15 | 9 | 144 | 270 | 306 | 18x(8, 15, 17) |
| 15 | 10 | 125 | 300 | 325 | 25x(5, 12, 13) |
| 15 | 11 | 104 | 330 | 346 | 2x(52, 165, 173) |
| 15 | 12 | 81 | 360 | 363 | 9x(9, 40, 41) |
| 15 | 13 | 56 | 390 | 394 | 2x(28, 195. 197) |
| 15 | 14 | 29 | 420 | 421 | Primitive |
| 16 | 1 | 255 | 32 | 257 | Primitive |
| 16 | 2 | 252 | 64 | 260 | 4x(63, 16, 65) |
| 16 | 3 | 247 | 96 | 265 | Primitive |
| 16 | 4 | 240 | 128 | 272 | 16x(15, 8, 17) |
| 16 | 5 | 231 | 160 | 281 | Primitive |
| 16 | 6 | 220 | 192 | 292 | 4x(55, 48, 73) |
| 16 | 7 | 207 | 224 | 305 | Primitive |
| 16 | 8 | 192 | 256 | 320 | 64x(3, 4, 5) |
| 16 | 9 | 175 | 288 | 337 | Primitive |
| 16 | 10 | 156 | 320 | 356 | 4x(39, 80, 89) |
| 16 | 11 | 135 | 352 | 377 | Primitive |
| 16 | 12 | 112 | 384 | 400 | 16x(7, 24, 25) |
| 16 | 13 | 87 | 416 | 425 | Primitive |
| 16 | 14 | 60 | 448 | 452 | 4x(15, 112, 113) |
| 16 | 15 | 31 | 480 | 481 | Primitive |
| 17 | 1 | 288 | 34 | 290 | 2x(144, 17, 145) |
| 17 | 2 | 285 | 68 | 293 | Primitive |
| 17 | 3 | 280 | 102 | 298 | 2x(140. 51. 149) |
| 17 | 4 | 273 | 136 | 305 | Primitive |
| 17 | 5 | 264 | 170 | 314 | 2x(132. 85, 157) |
| 17 | 6 | 253 | 204 | 325 | Primitive |
| 17 | 7 | 240 | 238 | 338 | 2x(12. 119. 169) |
| 17 | 8 | 225 | 272 | 353 | Primitive |
| 17 | 9 | 208 | 306 | 370 | 2x(104, 153. 185) |
| 17 | 10 | 189 | 340 | 389 | Primitive |
| 17 | 11 | 168 | 374 | 410 | 2x(84, 187, 205) |
| 17 | 12 | 145 | 408 | 433 | Primitive |
| 17 | 13 | 120 | 442 | 458 | 2x(60. 221. 229) |
| 17 | 14 | 93 | 476 | 485 | Primitive |
| 17 | 15 | 64 | 510 | 514 | 2x(32. 255. 257) |
| 17 | 16 | 33 | 544 | 545 | Primitive |
| 18 | 1 | 323 | 36 | 325 | Primitive |
| 18 | 2 | 320 | 72 | 328 | 8x(40. 9. 41) |
| 18 | 3 | 315 | 108 | 333 | 9x(35. 12. 37) |
| 18 | 4 | 308 | 144 | 340 | 4x(77. 36. 85) |
| 18 | 5 | 299 | 180 | 349 | Primitive |
| 18 | 6 | 288 | 216 | 360 | 72x(4. 3, 5) |
| 18 | 7 | 275 | 252 | 373 | Primitive |
| 18 | 8 | 260 | 288 | 388 | 4x(65, 72, 97) |

| $m$ | $n$ | $a = m^2 - n^2$ | $b = 2mn$ | $c = m^2 + n^2$ | *Primitive or multiple* |
|---|---|---|---|---|---|
| 18 | 9 | 243 | 324 | 405 | 81x(3, 4, 5) |
| 18 | 10 | 224 | 360 | 424 | 8x(28, 45, 53) |
| 18 | 11 | 203 | 396 | 445 | Primitive |
| 18 | 12 | 180 | 432 | 468 | 36x(5, 12, 13) |
| 18 | 13 | 155 | 468 | 493 | Primitive |
| 18 | 14 | 128 | 504 | 520 | 8x(16, 63, 65) |
| 18 | 15 | 99 | 540 | 549 | 9x(11, 60, 61) |
| 18 | 16 | 68 | 576 | 580 | 4x(17, 144, 145) |
| 18 | 17 | 35 | 612 | 613 | Primitive |
| 19 | 1 | 360 | 38 | 362 | 2x(180, 19, 181) |
| 19 | 2 | 357 | 76 | 365 | Primitive |
| 19 | 3 | 352 | 114 | 370 | 2x(176, 57, 185) |
| 19 | 4 | 345 | 152 | 377 | Primitive |
| 19 | 5 | 336 | 190 | 386 | 2x(168, 95, 193) |
| 19 | 6 | 325 | 228 | 397 | Primitive |
| 19 | 7 | 312 | 266 | 410 | 2x(156, 133, 205) |
| 19 | 8 | 297 | 304 | 425 | Primitive |
| 19 | 9 | 280 | 342 | 442 | 2x(140, 171, 221) |
| 19 | 10 | 261 | 380 | 461 | Primitive |
| 19 | 11 | 240 | 418 | 482 | 2x(120, 209, 241) |
| 19 | 12 | 217 | 456 | 505 | Primitive |
| 19 | 13 | 192 | 494 | 530 | 2x(96, 247, 265) |
| 19 | 14 | 165 | 532 | 557 | Primitive |
| 19 | 15 | 136 | 570 | 586 | 2x(68, 285, 293) |
| 19 | 16 | 105 | 608 | 617 | Primitive |
| 19 | 17 | 72 | 646 | 650 | 2x(36, 323, 325) |
| 19 | 18 | 37 | 684 | 685 | Primitive |
| 20 | 1 | 399 | 40 | 401 | Primitive |
| 20 | 2 | 396 | 80 | 404 | 4x(99, 20, 101) |
| 20 | 3 | 391 | 120 | 409 | Primitive |
| 20 | 4 | 384 | 160 | 416 | 32x(12, 5, 13) |
| 20 | 5 | 375 | 200 | 425 | 25x(15, 8, 17) |
| 20 | 6 | 364 | 240 | 436 | 4x(91, 60, 109) |
| 20 | 7 | 351 | 280 | 449 | Primitive |
| 20 | 8 | 336 | 320 | 464 | 16x(21, 20, 29) |
| 20 | 9 | 319 | 360 | 481 | Primitive |
| 20 | 10 | 300 | 400 | 500 | 100x(3, 4, 5) |
| 20 | 11 | 279 | 440 | 521 | Primitive |
| 20 | 12 | 256 | 480 | 544 | 32x(8, 15, 17) |
| 20 | 13 | 231 | 52 | 569 | Primitive |
| 20 | 14 | 204 | 560 | 596 | 4x(51, 140, 149) |
| 20 | 15 | 175 | 600 | 625 | 25x(7, 24, 25) |
| 20 | 16 | 144 | 640 | 656 | 16x(9, 40, 41) |
| 20 | 17 | 111 | 680 | 689 | Primitive |
| 20 | 18 | 76 | 720 | 724 | 4x(19, 180, 181) |
| 20 | 19 | 39 | 760 | 761 | Primitive |
| 21 | 1 | 440 | 42 | 442 | 2x(220, 21, 221) |
| 21 | 2 | 437 | 84 | 445 | Primitive |
| 21 | 3 | 432 | 126 | 450 | 18x(24, 7, 25) |
| 21 | 4 | 425 | 168 | 457 | Primitive |
| 21 | 5 | 416 | 210 | 466 | 2x(208, 105, 233) |
| 21 | 6 | 405 | 252 | 477 | 9x(45, 28, 53) |
| 21 | 7 | 392 | 294 | 490 | 98x(4, 3, 5) |
| 21 | 8 | 377 | 336 | 505 | Primitive |
| 21 | 9 | 360 | 378 | 522 | 18x(20, 21, 29) |
| 21 | 10 | 341 | 420 | 541 | Primitive |

| $m$ | $n$ | $a = m^2 - n^2$ | $b = 2mn$ | $c = m^2 + n^2$ | *Primitive or multiple* |
|---|---|---|---|---|---|
| 21 | 11 | 320 | 462 | 562 | 2x(160, 231, 281) |
| 21 | 12 | 297 | 504 | 585 | 9x(33, 56, 65) |
| 21 | 13 | 272 | 546 | 610 | 2x(136, 273, 305) |
| 21 | 14 | 245 | 588 | 637 | 49x(5, 12, 13) |
| 21 | 15 | 216 | 630 | 666 | 18x(12, 35, 37) |
| 21 | 16 | 185 | 672 | 697 | Primitive |
| 21 | 17 | 152 | 714 | 730 | 2x(76, 357, 365) |
| 21 | 18 | 117 | 756 | 765 | 9x(13, 84, 85) |
| 21 | 19 | 80 | 798 | 802 | 2x(40, 399, 401) |
| 21 | 20 | 41 | 840 | 841 | Primitive |
| 22 | 1 | 483 | 44 | 485 | Primitive |
| 22 | 2 | 480 | 88 | 488 | 8x(60, 11, 61) |
| 22 | 3 | 475 | 132 | 493 | Primitive |
| 22 | 4 | 468 | 176 | 500 | 4x(117, 44, 125) |
| 22 | 5 | 459 | 220 | 509 | Primitive |
| 22 | 6 | 448 | 264 | 520 | 8x(56, 33, 65) |
| 22 | 7 | 435 | 308 | 533 | Primitive |
| 22 | 8 | 420 | 352 | 548 | 4x(105, 88, 137) |
| 22 | 9 | 403 | 396 | 565 | Primitive |
| 22 | 10 | 384 | 440 | 584 | 8x(48, 55, 73) |
| 22 | 11 | 363 | 484 | 605 | 121x(3, 4, 5) |
| 22 | 12 | 340 | 528 | 628 | 4x(85, 132, 157) |
| 22 | 13 | 315 | 572 | 653 | Primitive |
| 22 | 14 | 288 | 616 | 680 | 8x(36, 77, 85) |
| 22 | 15 | 259 | 660 | 709 | Primitive |
| 22 | 16 | 228 | 704 | 740 | 4x(57, 176, 185) |
| 22 | 17 | 195 | 748 | 773 | Primitive |
| 22 | 18 | 160 | 792 | 808 | 8x(20, 99, 101) |
| 22 | 19 | 123 | 836 | 845 | Primitive |
| 22 | 20 | 84 | 880 | 884 | 4x(21, 220, 221) |
| 22 | 21 | 43 | 924 | 925 | Primitive |
| 23 | 1 | 528 | 46 | 530 | 2x(264, 23, 265) |
| 23 | 2 | 525 | 92 | 533 | Primitive |
| 23 | 3 | 520 | 138 | 538 | 2x(260, 69, 269) |
| 23 | 4 | 513 | 184 | 545 | Primitive |
| 23 | 5 | 504 | 230 | 554 | 2x(252, 115, 277) |
| 23 | 6 | 493 | 276 | 565 | Primitive |
| 23 | 7 | 480 | 322 | 578 | 2x(240, 161, 289) |
| 23 | 8 | 465 | 368 | 593 | Primitive |
| 23 | 9 | 448 | 414 | 610 | 2x(224, 207, 305) |
| 23 | 10 | 429 | 460 | 629 | Primitive |
| 23 | 11 | 408 | 506 | 650 | 2x(204, 253, 325) |
| 23 | 12 | 385 | 552 | 673 | Primitive |
| 23 | 13 | 360 | 598 | 698 | 2x(180, 299, 349) |
| 23 | 14 | 333 | 644 | 725 | Primitive |
| 23 | 15 | 304 | 690 | 754 | 2x(152, 345, 377) |
| 23 | 16 | 273 | 736 | 785 | Primitive |
| 23 | 17 | 240 | 782 | 818 | 2x(120, 391, 409) |
| 23 | 18 | 205 | 828 | 853 | Primitive |
| 23 | 19 | 168 | 874 | 890 | 2x(84, 437, 445) |
| 23 | 20 | 129 | 920 | 929 | Primitive |
| 23 | 21 | 88 | 966 | 970 | 2x(44, 483, 485) |
| 23 | 22 | 45 | 1012 | 1013 | Primitive |
| 24 | 1 | 575 | 48 | 577 | Primitive |
| 24 | 2 | 572 | 96 | 580 | 4x(143, 24, 145) |
| 24 | 3 | 567 | 144 | 585 | 9x(63, 16, 65) |

| $m$ | $n$ | $a = m^2 - n^2$ | $b = 2mn$ | $c = m^2 + n^2$ | *Primitive or multiple* |
|---|---|---|---|---|---|
| 24 | 4 | 560 | 192 | 592 | 16x(35, 12, 37) |
| 24 | 5 | 551 | 240 | 601 | Primitive |
| 24 | 6 | 540 | 288 | 612 | 36x(15, 8, 17) |
| 24 | 7 | 527 | 336 | 625 | Primitive |
| 24 | 8 | 512 | 384 | 640 | 128x(4, 3, 5) |
| 24 | 9 | 496 | 432 | 657 | 9x(55, 48, 73) |
| 24 | 10 | 476 | 480 | 676 | 4x(119, 120, 169) |
| 24 | 11 | 455 | 528 | 697 | Primitive |
| 24 | 12 | 432 | 576 | 720 | 144x(3, 4, 5) |
| 24 | 13 | 407 | 624 | 745 | Primitive |
| 24 | 14 | 380 | 672 | 772 | 4x(95, 168, 193) |
| 24 | 15 | 351 | 720 | 801 | 9x(39, 80, 89) |
| 24 | 16 | 320 | 768 | 832 | 64x(5, 12, 13) |
| 24 | 17 | 287 | 816 | 865 | Primitive |
| 24 | 18 | 252 | 864 | 900 | 36x(7, 24, 25) |
| 24 | 19 | 215 | 912 | 937 | Primitive |
| 24 | 20 | 176 | 960 | 976 | 16x(11, 60, 61) |
| 24 | 21 | 135 | 1008 | 1017 | 9x(15, 112, 113) |
| 24 | 22 | 92 | 1056 | 1060 | 4x(23, 264, 265) |
| 24 | 23 | 47 | 1104 | 1105 | Primitive |
| 25 | 1 | 624 | 50 | 626 | 2x(312, 25, 313) |
| 25 | 2 | 621 | 100 | 629 | Primitive |
| 25 | 3 | 616 | 150 | 634 | 2x(308, 75, 317) |
| 25 | 4 | 609 | 200 | 641 | Primitive |
| 25 | 5 | 600 | 250 | 650 | 50x(12, 5, 13) |
| 25 | 6 | 589 | 300 | 661 | Primitive |
| 25 | 7 | 576 | 350 | 674 | 2x(288, 175, 337) |
| 25 | 8 | 561 | 400 | 689 | Primitive |
| 25 | 9 | 544 | 450 | 706 | 2x(272, 225, 353) |
| 25 | 10 | 525 | 500 | 725 | 25x(21, 20, 29) |
| 25 | 11 | 504 | 550 | 746 | 2x(252, 275, 373) |
| 25 | 12 | 481 | 600 | 769 | Primitive |
| 25 | 13 | 456 | 650 | 794 | 2x(228, 325, 397) |
| 25 | 14 | 429 | 700 | 821 | Primitive |
| 25 | 15 | 400 | 750 | 850 | 50x(8, 15, 17) |
| 25 | 16 | 369 | 800 | 881 | Primitive |
| 25 | 17 | 336 | 850 | 914 | 2x(168, 425, 457) |
| 25 | 18 | 301 | 900 | 949 | Primitive |
| 25 | 19 | 264 | 950 | 986 | 2x(132, 475, 493) |
| 25 | 20 | 225 | 1000 | 1025 | 25x(9, 40, 41) |
| 25 | 21 | 184 | 1050 | 1066 | 2x(92, 525, 533) |
| 25 | 22 | 141 | 1100 | 1109 | Primitive |
| 25 | 23 | 96 | 1150 | 1154 | 2x(48, 575, 577) |
| 25 | 24 | 49 | 1200 | 1201 | Primitive |
| 26 | 1 | 675 | 52 | 677 | Primitive |
| 26 | 2 | 672 | 104 | 680 | 8x(84, 13, 85) |
| 26 | 3 | 667 | 156 | 685 | Primitive |
| 26 | 4 | 660 | 208 | 692 | 4x(165, 52, 173) |
| 26 | 5 | 651 | 260 | 701 | Primitive |
| 26 | 6 | 640 | 312 | 712 | 8x(80, 39, 89) |
| 26 | 7 | 627 | 364 | 725 | Primitive |
| 26 | 8 | 612 | 416 | 740 | 4x(153, 104, 185) |
| 26 | 9 | 595 | 468 | 757 | Primitive |
| 26 | 10 | 576 | 520 | 776 | 8x(72, 65, 97) |
| 26 | 11 | 555 | 572 | 797 | Primitive |
| 26 | 12 | 532 | 624 | 820 | 4x(133, 156, 205) |

| $m$ | $n$ | $a = m^2 - n^2$ | $b = 2mn$ | $c = m^2 + n^2$ | *Primitive or multiple* |
|---|---|---|---|---|---|
| 26 | 13 | 507 | 676 | 845 | 169x(3, 4, 5) |
| 26 | 14 | 480 | 728 | 872 | 8x(60, 91, 109) |
| 26 | 15 | 451 | 780 | 901 | Primitive |
| 26 | 16 | 420 | 832 | 932 | 4x(105, 208, 233) |
| 26 | 17 | 387 | 884 | 965 | Primitive |
| 26 | 18 | 352 | 936 | 1000 | 8x(44, 117, 125) |
| 26 | 19 | 315 | 988 | 1037 | Primitive |
| 26 | 20 | 276 | 1040 | 1076 | 4x(69, 260, 269) |
| 26 | 21 | 235 | 1092 | 1117 | Primitive |
| 26 | 22 | 192 | 1144 | 1160 | 8x(24, 143, 145) |
| 26 | 23 | 147 | 1196 | 1205 | Primitive |
| 26 | 24 | 100 | 1248 | 1252 | 4x(25, 312, 313) |
| 26 | 25 | 51 | 1300 | 1301 | Primitive |
| 27 | 1 | 728 | 54 | 730 | 2x(364, 27, 365) |
| 27 | 2 | 725 | 108 | 733 | Primitive |
| 27 | 3 | 720 | 162 | 738 | 18x(40, 9, 41) |
| 27 | 4 | 713 | 216 | 745 | Primitive |
| 27 | 5 | 704 | 270 | 754 | 2x(352, 135, 377) |
| 27 | 6 | 693 | 324 | 765 | 9x(77, 36, 85) |
| 27 | 7 | 680 | 378 | 778 | 2x(340, 189, 389) |
| 27 | 8 | 665 | 432 | 793 | Primitive |
| 27 | 9 | 648 | 486 | 810 | 162x(4, 3, 5) |
| 27 | 10 | 629 | 540 | 829 | Primitive |
| 27 | 11 | 608 | 594 | 850 | 2x(304, 297, 425) |
| 27 | 12 | 585 | 648 | 873 | 9x(65, 72, 97) |
| 27 | 13 | 560 | 702 | 898 | 2x(280, 351, 449) |
| 27 | 14 | 533 | 756 | 925 | Primitive |
| 27 | 15 | 504 | 810 | 954 | 18x(28, 45, 53) |
| 27 | 16 | 473 | 864 | 985 | Primitive |
| 27 | 17 | 440 | 918 | 1018 | 2x(220, 459, 509) |
| 27 | 18 | 405 | 972 | 1053 | 81x(5, 12, 13) |
| 27 | 19 | 368 | 1026 | 1090 | 2x(184, 513, 545) |
| 27 | 20 | 329 | 1080 | 1129 | Primitive |
| 27 | 21 | 288 | 1134 | 1170 | 18x(16, 63, 65) |
| 27 | 22 | 245 | 1188 | 1213 | Primitive |
| 27 | 23 | 200 | 1242 | 1258 | 2x(100, 621, 629) |
| 27 | 24 | 153 | 1296 | 1305 | 9x(17, 144, 145) |
| 27 | 25 | 104 | 1350 | 1354 | 2x(52, 675, 677) |
| 27 | 26 | 53 | 1404 | 1405 | Primitive |
| 28 | 1 | 783 | 56 | 785 | Primitive |
| 28 | 2 | 780 | 112 | 788 | 4x(195, 28, 197) |
| 28 | 3 | 775 | 168 | 793 | Primitive |
| 28 | 4 | 768 | 224 | 800 | 32x(24, 7, 25) |
| 28 | 5 | 759 | 280 | 809 | Primitive |
| 28 | 6 | 748 | 336 | 820 | 4x(187, 84, 205) |
| 28 | 7 | 735 | 392 | 833 | 49x(15, 8, 17) |
| 28 | 8 | 720 | 448 | 848 | 16x(45, 28, 53) |
| 28 | 9 | 703 | 504 | 865 | Primitive |
| 28 | 10 | 684 | 560 | 884 | 4x(171, 140, 221) |
| 28 | 11 | 663 | 616 | 905 | Primitive |
| 28 | 12 | 640 | 672 | 928 | 32x(20, 21, 29) |
| 28 | 13 | 615 | 728 | 953 | Primitive |
| 28 | 14 | 588 | 784 | 980 | 196x(3, 4, 5) |
| 28 | 15 | 559 | 840 | 1009 | Primitive |
| 28 | 16 | 528 | 896 | 1040 | 16x(33, 56, 65) |
| 28 | 17 | 495 | 952 | 1073 | Primitive |

| $m$ | $n$ | $a = m^2 - n^2$ | $b = 2mn$ | $c = m^2 + n^2$ | *Primitive or multiple* |
|---|---|---|---|---|---|
| 28 | 18 | 460 | 1008 | 1108 | 4x(115, 252, 277) |
| 28 | 19 | 423 | 1064 | 1145 | Primitive |
| 28 | 20 | 384 | 1120 | 1184 | 32x(12, 35, 37) |
| 28 | 21 | 343 | 1176 | 1225 | 49x(7, 24, 25) |
| 28 | 22 | 300 | 1232 | 1268 | 4x(75, 308, 317) |
| 28 | 23 | 255 | 1288 | 1313 | Primitive |
| 28 | 24 | 208 | 1344 | 1360 | 16x(13, 84, 85) |
| 28 | 25 | 159 | 1400 | 1409 | Primitive |
| 28 | 26 | 108 | 1456 | 1460 | 4x(27, 364, 365) |
| 28 | 27 | 55 | 1512 | 1513 | Primitive |
| 29 | 1 | 840 | 58 | 842 | 2x(420, 29, 421) |
| 29 | 2 | 837 | 116 | 845 | Primitive |
| 29 | 3 | 832 | 174 | 850 | 2x(416, 87, 425) |
| 29 | 4 | 825 | 232 | 857 | Primitive |
| 29 | 5 | 816 | 290 | 866 | 2x(408, 145, 433) |
| 29 | 6 | 805 | 348 | 877 | Primitive |
| 29 | 7 | 792 | 406 | 890 | 2x(396, 203, 445) |
| 29 | 8 | 777 | 464 | 905 | Primitive |
| 29 | 9 | 760 | 522 | 922 | 2x(380, 261, 461) |
| 29 | 10 | 741 | 580 | 941 | Primitive |
| 29 | 11 | 720 | 638 | 962 | 2x(360, 319, 481) |
| 29 | 12 | 697 | 696 | 985 | Primitive |
| 29 | 13 | 672 | 754 | 1010 | 2x(336, 377, 505) |
| 29 | 14 | 645 | 812 | 1037 | Primitive |
| 29 | 15 | 616 | 870 | 1066 | 2x(308, 435, 533) |
| 29 | 16 | 585 | 928 | 1097 | Primitive |
| 29 | 17 | 552 | 986 | 1130 | 2x(276, 493, 565) |
| 29 | 18 | 517 | 1044 | 1165 | Primitive |
| 29 | 19 | 480 | 1102 | 1202 | 2x(240, 551, 601) |
| 29 | 20 | 441 | 1160 | 1241 | Primitive |
| 29 | 21 | 400 | 1218 | 1282 | 2x(200, 609, 641) |
| 29 | 22 | 357 | 1276 | 1325 | Primitive |
| 29 | 23 | 312 | 1334 | 1370 | 2x(156, 667, 685) |
| 29 | 24 | 265 | 1392 | 1417 | Primitive |
| 29 | 25 | 216 | 1450 | 1466 | 2x(108, 725, 733) |
| 29 | 26 | 165 | 1508 | 1517 | Primitive |
| 29 | 27 | 112 | 1566 | 1570 | 2x(56, 783, 785) |
| 29 | 28 | 57 | 1624 | 1625 | Primitive |
| 30 | 1 | 899 | 60 | 901 | Primitive |
| 30 | 2 | 896 | 120 | 904 | 8x(112, 15, 113) |
| 30 | 3 | 891 | 180 | 909 | 9x(99, 20, 101) |
| 30 | 4 | 884 | 240 | 916 | 4x(221, 60, 229) |
| 30 | 5 | 875 | 300 | 925 | 25x(35, 12, 37) |
| 30 | 6 | 864 | 360 | 936 | 72x(12, 5, 13) |
| 30 | 7 | 851 | 420 | 949 | Primitive |
| 30 | 8 | 836 | 480 | 964 | 4x(209, 120, 241) |
| 30 | 9 | 819 | 540 | 981 | 9x(91, 60, 109) |
| 30 | 10 | 800 | 600 | 1000 | 200x(4, 3, 5) |
| 30 | 11 | 779 | 660 | 1021 | Primitive |
| 30 | 12 | 756 | 720 | 1044 | 36x(21, 20, 29) |
| 30 | 13 | 731 | 780 | 1069 | Primitive |
| 30 | 14 | 704 | 840 | 1096 | 8x(88, 105, 137) |
| 30 | 15 | 675 | 900 | 1125 | 225x(3, 4, 5) |
| 30 | 16 | 644 | 960 | 1156 | 4x(161, 240, 289) |
| 30 | 17 | 611 | 1020 | 1189 | Primitive |
| 30 | 18 | 576 | 1080 | 1224 | 72x(8, 15, 17) |

| $m$ | $n$ | $a = m^2 - n^2$ | $b = 2mn$ | $c = m^2 + n^2$ | *Primitive or multiple* |
|---|---|---|---|---|---|
| 30 | 19 | 539 | 1140 | 1261 | Primitive |
| 30 | 20 | 500 | 1200 | 1300 | 100x(5, 12, 13) |
| 30 | 21 | 459 | 1260 | 1341 | 9x(51, 140, 149) |
| 30 | 22 | 416 | 1320 | 1384 | 8x(52, 165, 173) |
| 30 | 23 | 371 | 1380 | 1429 | Primitive |
| 30 | 24 | 324 | 1440 | 1476 | 36x(9, 40, 41) |
| 30 | 25 | 275 | 1500 | 1525 | 25x(11, 60, 61) |
| 30 | 26 | 224 | 1560 | 1576 | 8x(28, 195, 197) |
| 30 | 27 | 171 | 1620 | 1629 | 9x(19, 180, 181) |
| 30 | 28 | 116 | 1680 | 1684 | 4x(29, 420, 421) |
| 30 | 29 | 59 | 1740 | 1741 | Primitive |
| 31 | 1 | 960 | 62 | 962 | 2x(480, 31, 481) |
| 31 | 2 | 957 | 124 | 965 | Primitive |
| 31 | 3 | 952 | 186 | 970 | 2x(476, 93, 485) |
| 31 | 4 | 945 | 248 | 977 | Primitive |
| 31 | 5 | 936 | 310 | 986 | 2x(468, 155. 493) |
| 31 | 6 | 925 | 372 | 997 | Primitive |
| 31 | 7 | 912 | 434 | 1010 | 2x(456, 217, 505) |
| 31 | 8 | 897 | 496 | 1025 | Primitive |
| 31 | 9 | 880 | 558 | 1042 | 2x(440, 279, 521) |
| 31 | 10 | 861 | 620 | 1061 | Primitive |
| 31 | 11 | 840 | 682 | 1082 | 2x(420, 341, 541) |
| 31 | 12 | 817 | 744 | 1105 | Primitive |
| 31 | 13 | 792 | 806 | 1130 | 2x(396, 403, 565) |
| 31 | 14 | 765 | 868 | 1157 | Primitive |
| 31 | 15 | 736 | 930 | 1186 | 2x(368, 465, 593) |
| 31 | 16 | 705 | 992 | 1217 | Primitive |
| 31 | 17 | 672 | 1054 | 1250 | 2x(336, 527, 625) |
| 31 | 18 | 637 | 1116 | 1285 | Primitive |
| 31 | 19 | 600 | 1178 | 1322 | 2x(300, 589, 661) |
| 31 | 20 | 561 | 1240 | 1361 | Primitive |
| 31 | 21 | 520 | 1302 | 1402 | 2x(260, 651, 701) |
| 31 | 22 | 477 | 1364 | 1445 | Primitive |
| 31 | 23 | 432 | 1426 | 1490 | 2x(216, 713, 745) |
| 31 | 24 | 385 | 1488 | 1537 | Primitive |
| 31 | 25 | 336 | 1550 | 1586 | 2x(168, 775, 793) |
| 31 | 26 | 285 | 1612 | 1637 | Primitive |
| 31 | 27 | 232 | 1674 | 1690 | 2x(116, 837, 845) |
| 31 | 28 | 177 | 1736 | 1745 | Primitive |
| 31 | 29 | 120 | 1798 | 1802 | 2x(60, 899. 901) |
| 31 | 30 | 61 | 1860 | 1861 | Primitive |
| 32 | 1 | 1023 | 64 | 1025 | Primitive |
| 32 | 2 | 1020 | 128 | 1028 | 4x(255, 32, 257) |
| 32 | 3 | 1015 | 192 | 1033 | Primitive |
| 32 | 4 | 1008 | 256 | 1040 | 16x(63. 16, 65) |
| 32 | 5 | 999 | 320 | 1049 | Primitive |
| 32 | 6 | 988 | 384 | 1060 | 4x(247, 96. 265) |
| 32 | 7 | 975 | 448 | 1073 | Primitive |
| 32 | 8 | 960 | 512 | 1088 | 64x(15, 8, 7) |
| 32 | 9 | 943 | 576 | 1105 | Primitive |
| 32 | 10 | 924 | 640 | 1124 | 4x(231. 160. 281) |
| 32 | 11 | 903 | 704 | 1145 | Primitive |
| 32 | 12 | 880 | 768 | 1168 | 16x(55. 48, 73) |
| 32 | 13 | 855 | 832 | 1193 | Primitive |
| 32 | 14 | 828 | 896 | 1220 | 4x(207, 224, 305) |
| 32 | 15 | 799 | 960 | 1249 | Primitive |

| $m$ | $n$ | $a = m^2 - n^2$ | $b = 2mn$ | $c = m^2 + n^2$ | *Primitive or multiple* |
|---|---|---|---|---|---|
| 32 | 16 | 768 | 1024 | 1280 | 256x(3, 4, 5) |
| 32 | 17 | 735 | 1088 | 1313 | Primitive |
| 32 | 18 | 700 | 1152 | 1348 | 4x(175, 288, 337) |
| 32 | 19 | 663 | 1216 | 1385 | Primitive |
| 32 | 20 | 624 | 1280 | 1424 | 16x(39, 80, 89) |
| 32 | 21 | 583 | 1344 | 1465 | Primitive |
| 32 | 22 | 540 | 1408 | 1508 | 4x(135, 352, 377) |
| 32 | 23 | 495 | 1472 | 1553 | Primitive |
| 32 | 24 | 448 | 1536 | 1600 | 64x(7, 24, 25) |
| 32 | 25 | 399 | 1600 | 1649 | Primitive |
| 32 | 26 | 348 | 1664 | 1700 | 4x(87, 416, 425) |
| 32 | 27 | 295 | 1728 | 1753 | Primitive |
| 32 | 28 | 240 | 1792 | 1808 | 16x(15, 112, 113) |
| 32 | 29 | 183 | 1856 | 1865 | Primitive |
| 32 | 30 | 124 | 1920 | 1924 | 4x(31, 480, 481) |
| 32 | 31 | 63 | 1984 | 1985 | Primitive |
| 33 | 1 | 1088 | 66 | 1090 | 2x(544, 33, 545) |
| 33 | 2 | 1085 | 132 | 1093 | Primitive |
| 33 | 3 | 1080 | 198 | 1098 | 18x(60, 11, 61) |
| 33 | 4 | 1073 | 264 | 1105 | Primitive |
| 33 | 5 | 1064 | 330 | 1114 | 2x(532, 165, 557) |
| 33 | 6 | 1053 | 396 | 1125 | 9x(117, 44, 125) |
| 33 | 7 | 1040 | 462 | 1138 | 2x(520, 231, 569) |
| 33 | 8 | 1025 | 528 | 1153 | Primitive |
| 33 | 9 | 1008 | 594 | 1170 | 18x(56, 33, 65) |
| 33 | 10 | 989 | 660 | 1189 | Primitive |
| 33 | 11 | 968 | 726 | 1210 | 242x(4, 3, 5) |
| 33 | 12 | 945 | 792 | 1233 | 9x(105, 88, 137) |
| 33 | 13 | 920 | 858 | 1258 | 2x(460. 429, 629) |
| 33 | 14 | 893 | 924 | 1285 | Primitive |
| 33 | 15 | 864 | 990 | 1314 | 18x(48, 55, 73) |
| 33 | 16 | 833 | 1056 | 1345 | Primitive |
| 33 | 17 | 800 | 1122 | 1378 | 2x(400, 561, 689) |
| 33 | 18 | 765 | 1188 | 1413 | 9x(85, 132, 157) |
| 33 | 19 | 728 | 1254 | 1450 | 2x(364. 627. 725) |
| 33 | 20 | 689 | 1320 | 1489 | Primitive |
| 33 | 21 | 648 | 1386 | 1530 | 18x(36, 77. 85) |
| 33 | 22 | 605 | 1452 | 1573 | 121x(5, 12. 13) |
| 33 | 23 | 560 | 1518 | 1618 | 2x(280. 759. 809) |
| 33 | 24 | 513 | 1584 | 1665 | 9x(57, 176, 185) |
| 33 | 25 | 464 | 1650 | 1714 | 2x(232, 825, 857) |
| 33 | 26 | 413 | 1716 | 1765 | Primitive |
| 33 | 27 | 360 | 1782 | 1818 | 18x(20. 99. 101) |
| 33 | 28 | 305 | 1848 | 1873 | Primitive |
| 33 | 29 | 248 | 1914 | 1930 | 2x(124. 957. 965) |
| 33 | 30 | 189 | 1980 | 1989 | 9x(21. 220. 221) |
| 33 | 31 | 128 | 2046 | 2050 | 2x(64. 1023. 1025) |
| 33 | 32 | 65 | 2112 | 2113 | Primitive |
| 34 | 1 | 1155 | 68 | 1157 | Primitive |
| 34 | 2 | 1152 | 136 | 1160 | 8x(144. 17. 145) |
| 34 | 3 | 1147 | 204 | 1165 | Primitive |
| 34 | 4 | 1140 | 272 | 1172 | 4x(285, 68. 293) |
| 34 | 5 | 1131 | 340 | 1181 | Primitive |
| 34 | 6 | 1120 | 408 | 1192 | 8x(140. 51. 149) |
| 34 | 7 | 1107 | 476 | 1205 | Primitive |
| 34 | 8 | 1092 | 544 | 1220 | 4x(273. 136. 305) |

| $m$ | $n$ | $a = m^2 - n^2$ | $b = 2mn$ | $c = m^2 + n^2$ | *Primitive or multiple* |
|---|---|---|---|---|---|
| 34 | 9 | 1075 | 612 | 1237 | Primitive |
| 34 | 10 | 1056 | 680 | 1256 | 8x(132, 85, 157) |
| 34 | 11 | 1035 | 748 | 1277 | Primitive |
| 34 | 12 | 1012 | 816 | 1300 | 4x(253, 204, 325) |
| 34 | 13 | 987 | 884 | 1325 | Primitive |
| 34 | 14 | 960 | 952 | 1352 | 8x(120, 119, 169) |
| 34 | 15 | 931 | 1020 | 1381 | Primitive |
| 34 | 16 | 900 | 1088 | 1412 | 4x(225, 272, 353) |
| 34 | 17 | 867 | 1156 | 1445 | 289x(3, 4, 5) |
| 34 | 18 | 832 | 1224 | 1480 | 8x(104, 153, 185) |
| 34 | 19 | 795 | 1292 | 1517 | Primitive |
| 34 | 20 | 756 | 1360 | 1556 | 4x(189, 340, 389) |
| 34 | 21 | 715 | 1428 | 1597 | Primitive |
| 34 | 22 | 672 | 1496 | 1640 | 8x(84, 187, 205) |
| 34 | 23 | 627 | 1564 | 1685 | Primitive |
| 34 | 24 | 580 | 1632 | 1732 | 4x(145, 408, 433) |
| 34 | 25 | 531 | 1700 | 1781 | Primitive |
| 34 | 26 | 480 | 1768 | 1832 | 8x(60, 221, 229) |
| 34 | 27 | 427 | 1836 | 1885 | Primitive |
| 34 | 28 | 372 | 1904 | 1940 | 4x(93, 476, 485) |
| 34 | 29 | 315 | 1972 | 1997 | Primitive |
| 34 | 30 | 256 | 2040 | 2056 | 8x(32, 255, 257) |
| 34 | 31 | 195 | 2108 | 2117 | Primitive |
| 34 | 32 | 132 | 2176 | 2180 | 4x(33, 544,545) |
| 34 | 33 | 67 | 2244 | 2245 | Primitive |
| 35 | 1 | 1224 | 70 | 1226 | 2x(612, 35, 613) |
| 35 | 2 | 1221 | 140 | 1229 | Primitive |
| 35 | 3 | 1216 | 210 | 1234 | 2x(608, 105, 617) |
| 35 | 4 | 1209 | 280 | 1241 | Primitive |
| 35 | 5 | 1200 | 350 | 1250 | 50x(24, 7, 25) |
| 35 | 6 | 1189 | 420 | 1261 | Primitive |
| 35 | 7 | 1176 | 490 | 1274 | 98x(12, 5, 13) |
| 35 | 8 | 1161 | 560 | 1289 | Primitive |
| 35 | 9 | 1144 | 630 | 1306 | 2x(572, 315, 653) |
| 35 | 10 | 1125 | 700 | 1325 | 25x(45, 28, 53) |
| 35 | 11 | 1104 | 770 | 1346 | 2x(552, 385, 673) |
| 35 | 12 | 1081 | 840 | 1369 | Primitive |
| 35 | 13 | 1056 | 910 | 1394 | 2x(528, 455, 697) |
| 35 | 14 | 1029 | 980 | 1421 | 49x(21, 20, 29) |
| 35 | 15 | 1000 | 1050 | 1450 | 50x(20, 21, 29) |
| 35 | 16 | 969 | 1120 | 1481 | Primitive |
| 35 | 17 | 936 | 1190 | 1514 | 2x(468, 595, 757) |
| 35 | 18 | 901 | 1260 | 1549 | Primitive |
| 35 | 19 | 864 | 1330 | 1586 | 2x(432, 665, 793) |
| 35 | 20 | 825 | 1400 | 1625 | 25x(33, 56, 65) |
| 35 | 21 | 784 | 1470 | 1666 | 98x(8, 15, 17) |
| 35 | 22 | 741 | 1540 | 1709 | Primitive |
| 35 | 23 | 696 | 1610 | 1754 | 2x(348, 805, 877) |
| 35 | 24 | 649 | 1680 | 1801 | Primitive |
| 35 | 25 | 600 | 1750 | 1850 | 50x(12, 35, 37) |
| 35 | 26 | 549 | 1820 | 1901 | Primitive |
| 35 | 27 | 496 | 1890 | 1954 | 2x(248, 945, 977) |
| 35 | 28 | 441 | 1960 | 2009 | 49x(9, 40, 41) |
| 35 | 29 | 384 | 2030 | 2066 | 2x(192, 1015, 1033) |
| 35 | 30 | 325 | 2100 | 2125 | 25x(13, 84, 85) |
| 35 | 31 | 264 | 2170 | 2186 | 2x(132, 1085, 1093) |

| $m$ | $n$ | $a = m^2 - n^2$ | $b = 2mn$ | $c = m^2 + n^2$ | *Primitive or multiple* |
|---|---|---|---|---|---|
| 35 | 32 | 201 | 2240 | 2249 | Primitive |
| 35 | 33 | 136 | 2310 | 2314 | 2x(68, 1155, 1157) |
| 35 | 34 | 69 | 2380 | 2381 | Primitive |
| 36 | 1 | 1295 | 72 | 1297 | Primitive |
| 36 | 2 | 1292 | 144 | 1300 | 4x(323, 36, 325) |
| 36 | 3 | 1287 | 216 | 1305 | 9x(143, 24, 145) |
| 36 | 4 | 1280 | 288 | 1312 | 32x(40, 9, 41) |
| 36 | 5 | 1271 | 360 | 1321 | Primitive |
| 36 | 6 | 1260 | 432 | 1332 | 36x(35, 12, 37) |
| 36 | 7 | 1247 | 504 | 1345 | Primitive |
| 36 | 8 | 1232 | 576 | 1360 | 16x(77, 36, 85) |
| 36 | 9 | 1215 | 648 | 1377 | 81x(15, 8, 17) |
| 36 | 10 | 1196 | 720 | 1396 | 4x(299, 180, 349) |
| 36 | 11 | 1175 | 792 | 1417 | Primitive |
| 36 | 12 | 1152 | 864 | 1440 | 288x(4, 3, 5) |
| 36 | 13 | 1127 | 936 | 1465 | Primitive |
| 36 | 14 | 1100 | 1008 | 1492 | 4x(275, 252, 373) |
| 36 | 15 | 1071 | 1080 | 1521 | 9x(119, 120, 169) |
| 36 | 16 | 1040 | 1152 | 1552 | 16x(65, 72, 97) |
| 36 | 17 | 1007 | 1224 | 1585 | Primitive |
| 36 | 18 | 972 | 1296 | 1620 | 324x(3, 4, 5) |
| 36 | 19 | 935 | 1368 | 1657 | Primitive |
| 36 | 20 | 896 | 1440 | 1696 | 32x(28, 45, 53) |
| 36 | 21 | 855 | 1512 | 1737 | 9x(95, 168, 193) |
| 36 | 22 | 812 | 1584 | 1780 | 4x(203, 396, 445) |
| 36 | 23 | 767 | 1656 | 1825 | Primitive |
| 36 | 24 | 720 | 1728 | 1872 | 144x(5, 12, 13) |
| 36 | 25 | 671 | 1800 | 1921 | Primitive |
| 36 | 26 | 620 | 1872 | 1972 | 4x(155, 468, 493) |
| 36 | 27 | 567 | 1944 | 2025 | 81x(7, 24, 25) |
| 36 | 28 | 512 | 2016 | 2080 | 32x(16, 63, 65) |
| 36 | 29 | 455 | 2088 | 2137 | Primitive |
| 36 | 30 | 396 | 2160 | 2196 | 36x(11, 60, 61) |
| 36 | 31 | 335 | 2232 | 2257 | Primitive |
| 36 | 32 | 272 | 2304 | 2320 | 16x(17, 144, 145) |
| 36 | 33 | 207 | 2376 | 2385 | 9x(23, 264, 265) |
| 36 | 34 | 140 | 2448 | 2452 | 4x(35, 612, 613) |
| 36 | 35 | 71 | 2520 | 2521 | Primitive |
| 37 | 1 | 1368 | 74 | 1370 | 2x(684, 37, 685) |
| 37 | 2 | 1365 | 148 | 1373 | Primitive |
| 37 | 3 | 1360 | 222 | 1378 | 2x(680, 111, 689) |
| 37 | 4 | 1353 | 296 | 1385 | Primitive |
| 37 | 5 | 1344 | 370 | 1394 | 2x(672, 185, 697) |
| 37 | 6 | 1333 | 444 | 1405 | Primitive |
| 37 | 7 | 1320 | 518 | 1418 | 2x(660, 259, 709) |
| 37 | 8 | 1305 | 592 | 1433 | Primitive |
| 37 | 9 | 1288 | 666 | 1450 | 2x(644, 333, 725) |
| 37 | 10 | 1269 | 740 | 1469 | Primitive |
| 37 | 11 | 1248 | 814 | 1490 | 2x(624, 407, 745) |
| 37 | 12 | 1225 | 888 | 1513 | Primitive |
| 37 | 13 | 1200 | 962 | 1538 | 2x(600, 481. 769) |
| 37 | 14 | 1173 | 1036 | 1565 | Primitive |
| 37 | 15 | 1144 | 1110 | 1594 | 2x(572, 555, 797) |
| 37 | 16 | 1113 | 1184 | 1625 | Primitive |
| 37 | 17 | 1080 | 1258 | 1658 | 2x(540, 629, 929) |
| 37 | 18 | 1045 | 1332 | 1693 | Primitive |

| $m$ | $n$ | $a = m^2 - n^2$ | $b = 2mn$ | $c = m^2 + n^2$ | *Primitive or multiple* |
|---|---|---|---|---|---|
| 37 | 19 | 1008 | 1406 | 1730 | 2x(504, 703, 865) |
| 37 | 20 | 969 | 1480 | 1769 | Primitive |
| 37 | 21 | 928 | 1554 | 1810 | 2x(464, 777, 905) |
| 37 | 22 | 885 | 1628 | 1853 | Primitive |
| 37 | 23 | 840 | 1702 | 1898 | 2x(420, 851, 949) |
| 37 | 24 | 793 | 1776 | 1945 | Primitive |
| 37 | 25 | 744 | 1850 | 1994 | 2x(372, 925, 997) |
| 37 | 26 | 693 | 1924 | 2045 | Primitive |
| 37 | 27 | 640 | 1998 | 2098 | 2x(320, 999, 1049) |
| 37 | 28 | 585 | 2072 | 2153 | Primitive |
| 37 | 29 | 528 | 2146 | 2210 | 2x(264, 1073, 1105) |
| 37 | 30 | 469 | 2220 | 2269 | Primitive |
| 37 | 31 | 408 | 2294 | 2330 | 2x(204, 1147, 1165) |
| 37 | 32 | 345 | 2368 | 2393 | Primitive |
| 37 | 33 | 280 | 2442 | 2458 | 2x(140, 1221, 1229) |
| 37 | 34 | 213 | 2516 | 2525 | Primitive |
| 37 | 35 | 144 | 2590 | 2594 | 2x(72, 1295, 1297) |
| 37 | 36 | 73 | 2664 | 2665 | Primitive |
| 38 | 1 | 1443 | 76 | 1445 | Primitive |
| 38 | 2 | 1440 | 152 | 1448 | 8x(180, 19. 181) |
| 38 | 3 | 1435 | 228 | 1453 | Primitive |
| 38 | 4 | 1428 | 304 | 1460 | 4x(357, 76. 365) |
| 38 | 5 | 1419 | 380 | 1469 | Primitive |
| 38 | 6 | 1408 | 456 | 1480 | 8x(176, 57, 185) |
| 38 | 7 | 1395 | 532 | 1493 | Primitive |
| 38 | 8 | 1380 | 608 | 1508 | 4x(345, 152, 377) |
| 38 | 9 | 1363 | 684 | 1525 | Primitive |
| 38 | 10 | 1344 | 760 | 1544 | 8x(168, 95, 193) |
| 38 | 11 | 1323 | 836 | 1565 | Primitive |
| 38 | 12 | 1300 | 912 | 1588 | 4x(325, 228, 397) |
| 38 | 13 | 1275 | 988 | 1613 | Primitive |
| 38 | 14 | 1248 | 1064 | 1640 | 8x(156, 133. 205) |
| 38 | 15 | 1219 | 1140 | 1669 | Primitive |
| 38 | 16 | 1188 | 1216 | 1700 | 4x(297, 304, 425) |
| 38 | 17 | 1155 | 1292 | 1733 | Primitive |
| 38 | 18 | 1120 | 1368 | 1768 | 8x(140, 171. 221) |
| 38 | 19 | 1083 | 1444 | 1805 | 361x(3, 4, 5) |
| 38 | 20 | 1044 | 1520 | 1844 | 4x(261, 380. 461) |
| 38 | 21 | 1003 | 1596 | 1885 | Primitive |
| 38 | 22 | 960 | 1672 | 1928 | 8x(120. 209. 241) |
| 38 | 23 | 915 | 1748 | 1973 | Primitive |
| 38 | 24 | 868 | 1824 | 2020 | 4x(217, 456, 505) |
| 38 | 25 | 819 | 1900 | 2069 | Primitive |
| 38 | 26 | 768 | 1976 | 2120 | 8x(96, 247, 265) |
| 38 | 27 | 715 | 2052 | 2173 | Primitive |
| 38 | 28 | 660 | 2128 | 2228 | 4x(165, 532, 557) |
| 38 | 29 | 603 | 2204 | 2285 | Primitive |
| 38 | 30 | 554 | 2280 | 2344 | 8x(68, 285. 293) |
| 38 | 31 | 483 | 2356 | 2405 | Primitive |
| 38 | 32 | 420 | 2432 | 2468 | 4x(105. 608. 617) |
| 38 | 33 | 355 | 2508 | 2533 | Primitive |
| 38 | 34 | 288 | 2584 | 2600 | 8x(36. 323. 325) |
| 38 | 35 | 219 | 2660 | 2669 | Primitive |
| 38 | 36 | 148 | 2736 | 2740 | 4x(37. 684. 685) |
| 38 | 37 | 75 | 2812 | 2813 | Primitive |
| 39 | 1 | 1520 | 78 | 1522 | 2x(760. 39. 761) |

| $m$ | $n$ | $a = m^2 - n^2$ | $b = 2mn$ | $c = m^2 + n^2$ | *Primitive or multiple* |
|---|---|---|---|---|---|
| 39 | 2 | 1517 | 156 | 1525 | Primitive |
| 39 | 3 | 1512 | 234 | 1530 | 18x(84, 13, 85) |
| 39 | 4 | 1505 | 312 | 1537 | Primitive |
| 39 | 5 | 1496 | 390 | 1546 | 2x(748, 195, 773) |
| 39 | 6 | 1485 | 468 | 1557 | 9x(165, 52, 173) |
| 39 | 7 | 1472 | 546 | 1570 | 2x(736, 273, 785) |
| 39 | 8 | 1457 | 624 | 1585 | Primitive |
| 39 | 9 | 1440 | 702 | 1602 | 18x(80, 39, 89) |
| 39 | 10 | 1421 | 780 | 1621 | Primitive |
| 39 | 11 | 1400 | 858 | 1642 | 2x(700, 429, 821) |
| 39 | 12 | 1377 | 936 | 1665 | 9x(153, 104, 185) |
| 39 | 13 | 1352 | 1014 | 1690 | 338x(4, 3, 5) |
| 39 | 14 | 1325 | 1092 | 1717 | Primitive |
| 39 | 15 | 1296 | 1170 | 1746 | 18x(72, 65, 97) |
| 39 | 16 | 1265 | 1248 | 1777 | Primitive |
| 39 | 17 | 1232 | 1326 | 1810 | 2x(616, 663, 905) |
| 39 | 18 | 1197 | 1404 | 1845 | 9x(133, 156, 205) |
| 39 | 19 | 1160 | 1482 | 1882 | 2x(580, 741, 941) |
| 39 | 20 | 1121 | 1560 | 1921 | Primitive |
| 39 | 21 | 1080 | 1638 | 1962 | 18x(60, 91, 109) |
| 39 | 22 | 1037 | 1716 | 2005 | Primitive |
| 39 | 23 | 992 | 1794 | 2050 | 2x(496, 897, 1025) |
| 39 | 24 | 945 | 1872 | 2097 | 9x(105, 208, 233) |
| 39 | 25 | 896 | 1950 | 2146 | 2x(448, 975, 1073) |
| 39 | 26 | 845 | 2028 | 2197 | 169x(5, 12, 13) |
| 39 | 27 | 792 | 2106 | 2250 | 18x(44, 117, 125) |
| 39 | 28 | 737 | 2184 | 2305 | Primitive |
| 39 | 29 | 680 | 2262 | 2362 | 2x(340, 1131, 1181) |
| 39 | 30 | 621 | 2340 | 2421 | 9x(69, 260, 269) |
| 39 | 31 | 560 | 2418 | 2482 | 2x(280, 1209, 1241) |
| 39 | 32 | 497 | 2496 | 2545 | Primitive |
| 39 | 33 | 432 | 2574 | 2610 | 18x(24, 143, 145) |
| 39 | 34 | 365 | 2652 | 2677 | Primitive |
| 39 | 35 | 296 | 2730 | 2746 | 2x(148, 1365, 1373) |
| 39 | 36 | 225 | 2808 | 2817 | 9x(25. 312. 313) |
| 39 | 37 | 152 | 2886 | 2890 | 2x(76, 1443, 1445) |
| 39 | 38 | 77 | 2964 | 2965 | Primitive |
| 40 | 1 | 1599 | 80 | 1601 | Primitive |
| 40 | 2 | 1596 | 160 | 1604 | 4x(399 40, 401) |
| 40 | 3 | 1591 | 240 | 1609 | Primitive |
| 40 | 4 | 1584 | 320 | 1616 | 16x(99, 20, 101) |
| 40 | 5 | 1575 | 400 | 1625 | 25x(63, 16, 65) |
| 40 | 6 | 1564 | 480 | 1636 | 4x(391, 120, 409) |
| 40 | 7 | 1551 | 560 | 1649 | Primitive |
| 40 | 8 | 1536 | 640 | 1664 | 128x(12, 5, 13) |
| 40 | 9 | 1519 | 720 | 1681 | Primitive |
| 40 | 10 | 1500 | 800 | 1700 | 100x(15, 8, 17) |
| 40 | 11 | 1479 | 880 | 1721 | Primitive |
| 40 | 12 | 1456 | 960 | 1744 | 16x(91, 60, 109) |
| 40 | 13 | 1431 | 1040 | 1769 | Primitive |
| 40 | 14 | 1404 | 1120 | 1796 | 4x(351. 280, 449) |
| 40 | 15 | 1375 | 1200 | 1825 | 25x(55, 48, 73) |
| 40 | 16 | 1344 | 1280 | 1856 | 64x(21, 20, 29) |
| 40 | 17 | 1311 | 1360 | 1889 | Primitive |
| 40 | 18 | 1276 | 1440 | 1924 | 4x(319, 360, 481) |
| 40 | 19 | 1239 | 1520 | 1961 | Primitive |

| $m$ | $n$ | $a = m^2 - n^2$ | $b = 2mn$ | $c = m^2 + n^2$ | *Primitive or multiple* |
|---|---|---|---|---|---|
| 40 | 20 | 1200 | 1600 | 2000 | 400x(3, 4, 5) |
| 40 | 21 | 1159 | 1680 | 2041 | Primitive |
| 40 | 22 | 1116 | 1760 | 2084 | 4x(279, 440, 521) |
| 40 | 23 | 1071 | 1840 | 2129 | Primitive |
| 40 | 24 | 1024 | 1920 | 2176 | 128x(8, 15, 17) |
| 40 | 25 | 975 | 2000 | 2225 | 25x(39, 80, 89) |
| 40 | 26 | 924 | 2080 | 2276 | 4x(231, 520, 569) |
| 40 | 27 | 871 | 2160 | 2329 | Primitive |
| 40 | 28 | 816 | 2240 | 2384 | 16x(51, 140, 149) |
| 40 | 29 | 759 | 2320 | 2441 | Primitive |
| 40 | 30 | 700 | 2400 | 2500 | 100x(7, 24, 25) |
| 40 | 31 | 639 | 2480 | 2561 | Primitive |
| 40 | 32 | 576 | 2560 | 2624 | 64x(9, 40, 41) |
| 40 | 33 | 511 | 2640 | 2689 | Primitive |
| 40 | 34 | 444 | 2720 | 2756 | 4x(111, 680, 689) |
| 40 | 35 | 375 | 2800 | 2825 | 25x(15, 112. 113) |
| 40 | 36 | 304 | 2880 | 2896 | 16x(19, 180, 181) |
| 40 | 37 | 231 | 2960 | 2969 | Primitive |
| 40 | 38 | 156 | 3040 | 3044 | 4x(39, 760, 761) |
| 40 | 39 | 79 | 3120 | 3121 | Primitive |
| 41 | 1 | 1680 | 82 | 1682 | 2x(840, 41, 841) |
| 41 | 2 | 1677 | 164 | 1685 | Primitive |
| 41 | 3 | 1672 | 246 | 1690 | 2x(836, 123, 845) |
| 41 | 4 | 1665 | 328 | 1697 | Primitive |
| 41 | 5 | 1656 | 410 | 1706 | 2x(828, 205, 853) |
| 41 | 6 | 1645 | 492 | 1717 | Primitive |
| 41 | 7 | 1632 | 574 | 1730 | 2x(816. 287. 865) |
| 41 | 8 | 1617 | 656 | 1745 | Primitive |
| 41 | 9 | 1600 | 738 | 1762 | 2x(800. 369, 881) |
| 41 | 10 | 1581 | 820 | 1781 | Primitive |
| 41 | 11 | 1560 | 902 | 1802 | 2x(780, 451, 901) |
| 41 | 12 | 1537 | 984 | 1825 | Primitive |
| 41 | 13 | 1512 | 1066 | 1850 | 2x(756. 533, 925) |
| 41 | 14 | 1485 | 1148 | 1877 | Primitive |
| 41 | 15 | 1456 | 1230 | 1906 | 2x(728. 615. 953) |
| 41 | 16 | 1425 | 1312 | 1937 | Primitive |
| 41 | 17 | 1392 | 1394 | 1970 | 2x(696, 697, 985) |
| 41 | 18 | 1357 | 1476 | 2005 | Primitive |
| 41 | 19 | 1320 | 1558 | 2042 | 2x(660. 779. 1021) |
| 41 | 20 | 1281 | 1640 | 2081 | Primitive |
| 41 | 21 | 1240 | 1722 | 2122 | 2x(620, 861, 1061) |
| 41 | 22 | 1197 | 1804 | 2165 | Primitive |
| 41 | 23 | 1152 | 1886 | 2210 | 2x(576. 943, 1105) |
| 41 | 24 | 1105 | 1968 | 2257 | Primitive |
| 41 | 25 | 1056 | 2050 | 2306 | 2x(528. 1025. 1153) |
| 41 | 26 | 1005 | 2132 | 2357 | Primitive |
| 41 | 27 | 952 | 2214 | 2410 | 2x(476. 1107. 1205) |
| 41 | 28 | 897 | 2296 | 2465 | Primitive |
| 41 | 29 | 840 | 2378 | 2522 | 2x(420. 1189. 1261) |
| 41 | 30 | 781 | 2460 | 2581 | Primitive |
| 41 | 31 | 720 | 2542 | 2642 | 2x(360. 1271. 1321) |
| 41 | 32 | 657 | 2624 | 2705 | Primitive |
| 41 | 33 | 592 | 2706 | 2770 | 2x(296. 1353, 1385) |
| 41 | 34 | 525 | 2788 | 2837 | Primitive |
| 41 | 35 | 456 | 2870 | 2906 | 2x(228, 1435, 1453) |
| 41 | 36 | 385 | 2952 | 2977 | Primitive |

| $m$ | $n$ | $a = m^2 - n^2$ | $b = 2mn$ | $c = m^2 + n^2$ | *Primitive or multiple* |
|---|---|---|---|---|---|
| 41 | 37 | 312 | 3034 | 3050 | 2x(156, 1517, 1525) |
| 41 | 38 | 237 | 3116 | 3125 | Primitive |
| 41 | 39 | 160 | 3198 | 3202 | 2x(80, 1599, 1601) |
| 41 | 40 | 81 | 3280 | 3281 | Primitive |
| 42 | 1 | 1763 | 84 | 1765 | Primitive |
| 42 | 2 | 1760 | 168 | 1768 | 8x(220, 21, 221) |
| 42 | 3 | 1755 | 252 | 1773 | 9x(195, 28, 197) |
| 42 | 4 | 1748 | 336 | 1780 | 4x(437, 84, 445) |
| 42 | 5 | 1739 | 420 | 1789 | Primitive |
| 42 | 6 | 1728 | 504 | 1800 | 72x(24, 7, 25) |
| 42 | 7 | 1715 | 588 | 1813 | 49x(35, 12, 37) |
| 42 | 8 | 1700 | 672 | 1828 | 4x(425, 168, 457) |
| 42 | 9 | 1683 | 756 | 1845 | 9x(187, 84, 205) |
| 42 | 10 | 1664 | 840 | 1864 | 8x(208, 105, 233) |
| 42 | 11 | 1643 | 924 | 1885 | Primitive |
| 42 | 12 | 1620 | 1008 | 1908 | 36x(45. 28, 53) |
| 42 | 13 | 1595 | 1092 | 1933 | Primitive |
| 42 | 14 | 1568 | 1176 | 1960 | 392x(4, 3, 5) |
| 42 | 15 | 1539 | 1260 | 1989 | 9x(171, 140, 221) |
| 42 | 16 | 1508 | 1344 | 2020 | 4x(377, 336, 505) |
| 42 | 17 | 1475 | 1428 | 2053 | Primitive |
| 42 | 18 | 1440 | 1512 | 2088 | 72x(20, 21, 29) |
| 42 | 19 | 1403 | 1596 | 2125 | Primitive |
| 42 | 20 | 1364 | 1680 | 2164 | 4x(341, 420, 541) |
| 42 | 21 | 1323 | 1764 | 2205 | 441x(3, 4, 5) |
| 42 | 22 | 1280 | 1848 | 2248 | 8x(160, 231, 281) |
| 42 | 23 | 1235 | 1932 | 2293 | Primitive |
| 42 | 24 | 1188 | 2016 | 2340 | 36x(33, 56, 65) |
| 42 | 25 | 1139 | 2100 | 2389 | Primitive |
| 42 | 26 | 1088 | 2184 | 2440 | 8x(136, 273, 305) |
| 42 | 27 | 1035 | 2268 | 2493 | 9x(115, 252, 277) |
| 42 | 28 | 980 | 2352 | 2548 | 196x(5, 12, 13) |
| 42 | 29 | 923 | 2436 | 2605 | Primitive |
| 42 | 30 | 864 | 2520 | 2664 | 72x(12, 35, 37) |
| 42 | 31 | 803 | 2604 | 2725 | Primitive |
| 42 | 32 | 740 | 2688 | 2788 | 4x(185, 672, 697) |
| 42 | 33 | 675 | 2772 | 2853 | 9x(75, 308, 317) |
| 42 | 34 | 608 | 2856 | 2920 | 8x(76, 357, 365) |
| 42 | 35 | 539 | 2940 | 2989 | 49x(11, 60, 61) |
| 42 | 36 | 468 | 3024 | 3060 | 36x(13, 84, 85) |
| 42 | 37 | 395 | 3108 | 3133 | Primitive |
| 42 | 38 | 320 | 3192 | 3208 | 8x(40, 399, 401) |
| 42 | 39 | 243 | 3276 | 3285 | 9x(27, 364, 365) |
| 42 | 40 | 164 | 3360 | 3364 | 4x(41, 840, 841) |
| 42 | 41 | 83 | 3444 | 3445 | Primitive |
| 43 | 1 | 1848 | 86 | 1850 | 2x(924, 43. 925) |
| 43 | 2 | 1845 | 172 | 1853 | Primitive |
| 43 | 3 | 1840 | 258 | 1858 | 2x(920, 129, 929) |
| 43 | 4 | 1833 | 344 | 1865 | Primitive |
| 43 | 5 | 1824 | 430 | 1874 | 2x(912, 215, 937) |
| 43 | 6 | 1813 | 516 | 1885 | Primitive |
| 43 | 7 | 1800 | 602 | 1898 | 2x(900, 301, 949) |
| 43 | 8 | 1785 | 688 | 1913 | Primitive |
| 43 | 9 | 1768 | 774 | 1930 | 2x(884. 387. 965) |
| 43 | 10 | 1749 | 860 | 1949 | Primitive |
| 43 | 11 | 1728 | 946 | 1970 | 2x(864. 473. 985) |

| $m$ | $n$ | $a = m^2 - n^2$ | $b = 2mn$ | $c = m^2 + n^2$ | *Primitive or multiple* |
|---|---|---|---|---|---|
| 43 | 12 | 1705 | 1032 | 1993 | Primitive |
| 43 | 13 | 1680 | 1118 | 2018 | 2x(840, 559, 1009) |
| 43 | 14 | 1653 | 1204 | 2045 | Primitive |
| 43 | 15 | 1624 | 1290 | 2074 | 2x(812, 645, 1037) |
| 43 | 16 | 1593 | 1376 | 2105 | Primitive |
| 43 | 17 | 1560 | 1462 | 2138 | 2x(780, 731, 1069) |
| 43 | 18 | 1525 | 1548 | 2173 | Primitive |
| 43 | 19 | 1488 | 1634 | 2210 | 2x(744, 817, 1105) |
| 43 | 20 | 1449 | 1720 | 2249 | Primitive |
| 43 | 21 | 1408 | 1806 | 2290 | 2x(704, 903, 1145) |
| 43 | 22 | 1365 | 1892 | 2333 | Primitive |
| 43 | 23 | 1320 | 1978 | 2378 | 2x(660, 989, 1189) |
| 43 | 24 | 1273 | 2064 | 2425 | Primitive |
| 43 | 25 | 1224 | 2150 | 2474 | 2x(612, 1075, 1237) |
| 43 | 26 | 1173 | 2236 | 2525 | Primitive |
| 43 | 27 | 1120 | 2322 | 2578 | 2x(560, 1161, 1289) |
| 43 | 28 | 1065 | 2408 | 2633 | Primitive |
| 43 | 29 | 1008 | 2494 | 2690 | 2x(504, 1247, 1345) |
| 43 | 30 | 949 | 2580 | 2749 | Primitive |
| 43 | 31 | 888 | 2666 | 2810 | 2x(444, 1333. 1405) |
| 43 | 32 | 825 | 2752 | 2873 | Primitive |
| 43 | 33 | 760 | 2838 | 2939 | 2x(380, 1419. 1469) |
| 43 | 34 | 693 | 2924 | 3005 | Primitive |
| 43 | 35 | 624 | 3010 | 3074 | 2x(312, 1505, 1537) |
| 43 | 36 | 553 | 3096 | 3145 | Primitive |
| 43 | 37 | 480 | 3182 | 3218 | 2x(240, 1591, 1609) |
| 43 | 38 | 405 | 3268 | 3293 | Primitive |
| 43 | 39 | 328 | 3354 | 3370 | 2x(164, 1677, 1685) |
| 43 | 40 | 249 | 3440 | 3449 | Primitive |
| 43 | 41 | 168 | 3526 | 3530 | 2x(84, 1763, 1765) |
| 43 | 42 | 85 | 3612 | 3613 | Primitive |
| 44 | 1 | 1935 | 88 | 1937 | Primitive |
| 44 | 2 | 1932 | 176 | 1940 | 4x(483, 44, 485) |
| 44 | 3 | 1927 | 264 | 1945 | Primitive |
| 44 | 4 | 1920 | 352 | 1952 | 32x(60. 11. 61) |
| 44 | 5 | 1911 | 440 | 1961 | Primitive |
| 44 | 6 | 1900 | 528 | 1972 | 4x(475, 132, 493) |
| 44 | 7 | 1887 | 616 | 1985 | Primitive |
| 44 | 8 | 1872 | 704 | 2000 | 16x(117. 44. 125) |
| 44 | 9 | 1855 | 792 | 2017 | Primitive |
| 44 | 10 | 1836 | 880 | 2036 | 4x(459, 220, 509) |
| 44 | 11 | 1815 | 968 | 2057 | 121x(15, 8, 17) |
| 44 | 12 | 1792 | 1056 | 2080 | 32x(56, 33, 65) |
| 44 | 13 | 1767 | 1144 | 2105 | Primitive |
| 44 | 14 | 1740 | 1232 | 2132 | 4x(435, 308. 533) |
| 44 | 15 | 1711 | 1320 | 2161 | Primitive |
| 44 | 16 | 1680 | 1408 | 2192 | 16x(105. 88. 137) |
| 44 | 17 | 1647 | 1496 | 2225 | Primitive |
| 44 | 18 | 1612 | 1584 | 2260 | 4x(403, 396. 565) |
| 44 | 19 | 1575 | 1672 | 2297 | Primitive |
| 44 | 20 | 1536 | 1760 | 2336 | 32x(48. 55. 73) |
| 44 | 21 | 1495 | 1848 | 2377 | Primitive |
| 44 | 22 | 1452 | 1936 | 2420 | 484x(3, 4. 5) |
| 44 | 23 | 1407 | 2024 | 2465 | Primitive |
| 44 | 24 | 1360 | 2112 | 2512 | 16x(85, 132, 157) |
| 44 | 25 | 1311 | 2200 | 2561 | Primitive |

| $m$ | $n$ | $a = m^2 - n^2$ | $b = 2mn$ | $c = m^2 + n^2$ | *Primitive or multiple* |
|---|---|---|---|---|---|
| 44 | 26 | 1260 | 2288 | 2612 | 4x(315, 572, 653) |
| 44 | 27 | 1207 | 2376 | 2665 | Primitive |
| 44 | 28 | 1152 | 2464 | 2720 | 32x(36, 77, 85) |
| 44 | 29 | 1095 | 2552 | 2777 | Primitive |
| 44 | 30 | 1036 | 2640 | 2836 | 4x(259, 660, 709) |
| 44 | 31 | 975 | 2728 | 2897 | Primitive |
| 44 | 32 | 912 | 2816 | 2960 | 16x(57, 176, 185) |
| 44 | 33 | 847 | 2904 | 3025 | 121x(7, 24, 25) |
| 44 | 34 | 780 | 2992 | 3092 | 4x(195, 748, 773) |
| 44 | 35 | 711 | 3080 | 3161 | Primitive |
| 44 | 36 | 640 | 3168 | 3232 | 32x(20, 99, 101) |
| 44 | 37 | 567 | 3256 | 3305 | Primitive |
| 44 | 38 | 492 | 3344 | 3380 | 4x(123, 836, 845) |
| 44 | 39 | 415 | 3432 | 3457 | Primitive |
| 44 | 40 | 336 | 3520 | 3536 | 16x(21, 220, 221) |
| 44 | 41 | 255 | 3608 | 3617 | Primitive |
| 44 | 42 | 172 | 3696 | 3700 | 4x(43, 924, 925) |
| 44 | 43 | 87 | 3784 | 3785 | Primitive |
| 45 | 1 | 2024 | 90 | 2026 | 2x(1012, 45, 1013) |
| 45 | 2 | 2021 | 180 | 2029 | Primitive |
| 45 | 3 | 2016 | 270 | 2034 | 18x(112, 15, 113) |
| 45 | 4 | 2009 | 360 | 2041 | Primitive |
| 45 | 5 | 2000 | 450 | 2050 | 50x(40, 9, 41) |
| 45 | 6 | 1989 | 540 | 2061 | 9x(221, 60, 229) |
| 45 | 7 | 1976 | 630 | 2074 | 2x(988, 315, 1037) |
| 45 | 8 | 1961 | 720 | 2089 | Primitive |
| 45 | 9 | 1944 | 810 | 2106 | 162x(12, 5, 13) |
| 45 | 10 | 1925 | 900 | 2125 | 25x(77, 36, 85) |
| 45 | 11 | 1904 | 990 | 2146 | 2x(952, 495, 1073) |
| 45 | 12 | 1881 | 1080 | 2169 | 9x(209, 120, 241) |
| 45 | 13 | 1856 | 1170 | 2194 | 2x(928, 585, 1097) |
| 45 | 14 | 1829 | 1260 | 2221 | Primitive |
| 45 | 15 | 1800 | 1350 | 2250 | 450x(4, 3, 5) |
| 45 | 16 | 1769 | 1440 | 2281 | Primitive |
| 45 | 17 | 1736 | 1530 | 2314 | 2x(868, 765, 1157) |
| 45 | 18 | 1701 | 1620 | 2349 | 81x(21, 20, 29) |
| 45 | 19 | 1664 | 1710 | 2386 | 2x(832, 855, 1193) |
| 45 | 20 | 1625 | 1800 | 2425 | 25x(65, 72, 97) |
| 45 | 21 | 1584 | 1890 | 2466 | 18x(88, 105, 137) |
| 45 | 22 | 1541 | 1980 | 2509 | Primitive |
| 45 | 23 | 1496 | 2070 | 2554 | 2x(748, 1035, 1277) |
| 45 | 24 | 1449 | 2160 | 2601 | 9x(161, 240, 289) |
| 45 | 25 | 1400 | 2250 | 2650 | 50x(28, 45, 53) |
| 45 | 26 | 1349 | 2340 | 2701 | Primitive |
| 45 | 27 | 1296 | 2430 | 2754 | 162x(8, 15, 17) |
| 45 | 28 | 1241 | 2520 | 2809 | Primitive |
| 45 | 29 | 1184 | 2610 | 2866 | 2x(592, 1305, 1433) |
| 45 | 30 | 1125 | 2700 | 2925 | 225x(5, 12, 13) |
| 45 | 31 | 1064 | 2790 | 2986 | 2x(532, 1395, 1493) |
| 45 | 32 | 1001 | 2880 | 3049 | Primitive |
| 45 | 33 | 936 | 2970 | 3114 | 18x(52, 165, 173) |
| 45 | 34 | 869 | 3060 | 3181 | Primitive |
| 45 | 35 | 800 | 3150 | 3250 | 50x(16, 63, 65) |
| 45 | 36 | 729 | 3240 | 3321 | 81x(9, 40, 41) |
| 45 | 37 | 656 | 3330 | 3394 | 2x(328, 1665, 1697) |
| 45 | 38 | 581 | 3420 | 3469 | Primitive |

| $m$ | $n$ | $a = m^2 - n^2$ | $b = 2mn$ | $c = m^2 + n^2$ | *Primitive or multiple* |
|---|---|---|---|---|---|
| 45 | 39 | 504 | 3510 | 3546 | 18x(28, 195, 197) |
| 45 | 40 | 425 | 3600 | 3625 | 25x(17, 144, 145) |
| 45 | 41 | 344 | 3690 | 3706 | 2x(172, 1845, 1853) |
| 45 | 42 | 261 | 3780 | 3789 | 9x(29, 420, 421) |
| 45 | 43 | 176 | 3870 | 3874 | 2x(88, 1935, 1937) |
| 45 | 44 | 89 | 3960 | 3961 | Primitive |
| 46 | 1 | 2115 | 92 | 2117 | Primitive |
| 46 | 2 | 2112 | 184 | 2120 | 8x(264, 23, 265) |
| 46 | 3 | 2107 | 276 | 2125 | Primitive |
| 46 | 4 | 2100 | 368 | 2132 | 4x(525, 92, 533) |
| 46 | 5 | 2091 | 460 | 2141 | Primitive |
| 46 | 6 | 2080 | 552 | 2152 | 8x(260, 69. 269) |
| 46 | 7 | 2067 | 644 | 2165 | Primitive |
| 46 | 8 | 2052 | 736 | 2180 | 4x(513, 184. 545) |
| 46 | 9 | 2035 | 828 | 2197 | Primitive |
| 46 | 10 | 2016 | 920 | 2216 | 8x(252, 115. 277) |
| 46 | 11 | 1995 | 1012 | 2237 | Primitive |
| 46 | 12 | 1972 | 1104 | 2260 | 4x(493. 276, 565) |
| 46 | 13 | 1947 | 1196 | 2285 | Primitive |
| 46 | 14 | 1920 | 1288 | 2312 | 8x(240. 161. 289) |
| 46 | 15 | 1891 | 1380 | 2341 | Primitive |
| 46 | 16 | 1860 | 1472 | 2372 | 4x(465, 368. 593) |
| 46 | 17 | 1827 | 1564 | 2405 | Primitive |
| 46 | 18 | 1792 | 1656 | 2440 | 8x(224, 207, 305) |
| 46 | 19 | 1755 | 1748 | 2477 | Primitive |
| 46 | 20 | 1716 | 1840 | 2516 | 4x(429, 460, 629) |
| 46 | 21 | 1675 | 1932 | 2557 | Primitive |
| 46 | 22 | 1632 | 2024 | 2600 | 8x(204, 253. 325) |
| 46 | 23 | 1587 | 2116 | 2645 | Primitive |
| 46 | 24 | 1540 | 2208 | 2692 | 4x(385, 552, 673) |
| 46 | 25 | 1491 | 2300 | 2741 | Primitive |
| 46 | 26 | 1440 | 2392 | 2792 | 8x(180, 299. 349) |
| 46 | 27 | 1387 | 2484 | 2845 | Primitive |
| 46 | 28 | 1332 | 2576 | 2900 | 4x(333, 644, 725) |
| 46 | 29 | 1275 | 2668 | 2957 | Primitive |
| 46 | 30 | 1216 | 2760 | 3016 | 8x(152, 345, 377) |
| 46 | 31 | 1155 | 2852 | 3077 | Primitive |
| 46 | 32 | 1092 | 2944 | 3140 | 4x(273. 736, 785) |
| 46 | 33 | 1027 | 3036 | 3205 | Primitive |
| 46 | 34 | 960 | 3128 | 3272 | 8x(120. 391, 409) |
| 46 | 35 | 891 | 3220 | 3341 | Primitive |
| 46 | 36 | 820 | 3312 | 3412 | 4x(205, 828, 853) |
| 46 | 37 | 747 | 3404 | 3485 | Primitive |
| 46 | 38 | 672 | 3496 | 3560 | 8x(84, 437, 445) |
| 46 | 39 | 595 | 3588 | 3637 | Primitive |
| 46 | 40 | 516 | 3680 | 3716 | 4x(129, 920, 929) |
| 46 | 41 | 435 | 3772 | 3797 | Primitive |
| 46 | 42 | 352 | 3864 | 3880 | 8x(44, 483. 485) |
| 46 | 43 | 267 | 3956 | 3965 | Primitive |
| 46 | 44 | 180 | 4048 | 4052 | 4x(45. 1012. 1013) |
| 46 | 45 | 91 | 4140 | 4141 | Primitive |
| 47 | 1 | 2208 | 94 | 2210 | 2x(1104, 47. 1105) |
| 47 | 2 | 2205 | 188 | 2213 | Primitive |
| 47 | 3 | 2200 | 282 | 2218 | 2x(1100. 141. 1109) |
| 47 | 4 | 2193 | 376 | 2225 | Primitive |
| 47 | 5 | 2184 | 470 | 2234 | 2x(1092. 235, 1117) |

| $m$ | $n$ | $a = m^2 - n^2$ | $b = 2mn$ | $c = m^2 + n^2$ | *Primitive or multiple* |
|---|---|---|---|---|---|
| 47 | 6 | 2173 | 564 | 2245 | Primitive |
| 47 | 7 | 2160 | 658 | 2258 | 2x(1080, 329, 1129) |
| 47 | 8 | 2145 | 752 | 2273 | Primitive |
| 47 | 9 | 2128 | 846 | 2290 | 2x(1064, 423, 1145) |
| 47 | 10 | 2109 | 940 | 2309 | Primitive |
| 47 | 11 | 2088 | 1034 | 2330 | 2x(1044, 517, 1165) |
| 47 | 12 | 2065 | 1128 | 2353 | Primitive |
| 47 | 13 | 2040 | 1222 | 2378 | 2x(1020, 611, 1189) |
| 47 | 14 | 2013 | 1316 | 2405 | Primitive |
| 47 | 15 | 1984 | 1410 | 2434 | 2x(992, 705, 1217) |
| 47 | 16 | 1953 | 1504 | 2465 | Primitive |
| 47 | 17 | 1920 | 1598 | 2498 | 2x(960, 799, 1249) |
| 47 | 18 | 1885 | 1692 | 2533 | Primitive |
| 47 | 19 | 1848 | 1786 | 2570 | 2x(924, 893, 1285) |
| 47 | 20 | 1809 | 1880 | 2609 | Primitive |
| 47 | 21 | 1768 | 1974 | 2650 | 2x(884, 987, 1325) |
| 47 | 22 | 1725 | 2068 | 2693 | Primitive |
| 47 | 23 | 1680 | 2162 | 2738 | 2x(840, 1081, 1369) |
| 47 | 24 | 1633 | 2256 | 2785 | Primitive |
| 47 | 25 | 1584 | 2350 | 2834 | 2x(792, 1175, 1417) |
| 47 | 26 | 1533 | 2444 | 2885 | Primitive |
| 47 | 27 | 1480 | 2538 | 2938 | 2x(740, 1269, 1469) |
| 47 | 28 | 1425 | 2632 | 2993 | Primitive |
| 47 | 29 | 1368 | 2726 | 3050 | 2x(684, 1363, 1525) |
| 47 | 30 | 1309 | 2820 | 3109 | Primitive |
| 47 | 31 | 1248 | 2914 | 3170 | 2x(624, 1457, 1585) |
| 47 | 32 | 1185 | 3008 | 3233 | Primitive |
| 47 | 33 | 1120 | 3102 | 3298 | 2x(560, 1551, 1649) |
| 47 | 34 | 1053 | 3196 | 3365 | Primitive |
| 47 | 35 | 984 | 3290 | 3434 | 2x(492, 1645, 1717) |
| 47 | 36 | 913 | 3384 | 3505 | Primitive |
| 47 | 37 | 840 | 3478 | 3578 | 2x(420, 1739, 1789) |
| 47 | 38 | 765 | 3572 | 3653 | Primitive |
| 47 | 39 | 688 | 3666 | 3730 | 2x(344, 1833, 1865) |
| 47 | 40 | 609 | 3760 | 3809 | Primitive |
| 47 | 41 | 528 | 3854 | 3890 | 2x(264, 1927, 1945) |
| 47 | 42 | 445 | 3948 | 3973 | Primitive |
| 47 | 43 | 360 | 4042 | 4058 | 2x(180, 2021, 2029) |
| 47 | 44 | 273 | 4136 | 4145 | Primitive |
| 47 | 45 | 184 | 4230 | 4234 | 2x(92, 2115, 2117) |
| 47 | 46 | 93 | 4324 | 4325 | Primitive |
| 48 | 1 | 2303 | 96 | 2305 | Primitive |
| 48 | 2 | 2300 | 192 | 2308 | 4x(575, 48, 577) |
| 48 | 3 | 2295 | 288 | 2313 | 9x(255, 32, 257) |
| 48 | 4 | 2288 | 384 | 2320 | 16x(143, 24. 145) |
| 48 | 5 | 2279 | 480 | 2329 | Primitive |
| 48 | 6 | 2268 | 576 | 2340 | 36x(63. 16. 65) |
| 48 | 7 | 2255 | 672 | 2353 | Primitive |
| 48 | 8 | 2240 | 768 | 2368 | 64x(35, 12, 37) |
| 48 | 9 | 2223 | 864 | 2385 | 9x(247. 96. 265) |
| 48 | 10 | 2204 | 960 | 2404 | 4x(551. 240. 601) |
| 48 | 11 | 2183 | 1056 | 2425 | Primitive |
| 48 | 12 | 2160 | 1152 | 2448 | 144x(15, 8, 17) |
| 48 | 13 | 2135 | 1248 | 2473 | Primitive |
| 48 | 14 | 2108 | 1344 | 2500 | 4x(527, 336, 625) |
| 48 | 15 | 2079 | 1440 | 2529 | 9x(231. 160. 281) |

| $m$ | $n$ | $a = m^2 - n^2$ | $b = 2mn$ | $c = m^2 + n^2$ | *Primitive or multiple* |
|---|---|---|---|---|---|
| 48 | 16 | 2048 | 1536 | 2560 | 512x(4, 3, 5) |
| 48 | 17 | 2015 | 1632 | 2593 | Primitive |
| 48 | 18 | 1980 | 1728 | 2628 | 36x(55, 48, 73) |
| 48 | 19 | 1943 | 1824 | 2665 | Primitive |
| 48 | 20 | 1904 | 1920 | 2704 | 16x(119, 120, 169) |
| 48 | 21 | 1863 | 2016 | 2745 | 9x(207, 224, 305) |
| 48 | 22 | 1820 | 2112 | 2788 | 4x(455, 528, 697) |
| 48 | 23 | 1775 | 2208 | 2833 | Primitive |
| 48 | 24 | 1728 | 2304 | 2880 | 576x(3, 4, 5) |
| 48 | 25 | 1679 | 2400 | 2929 | Primitive |
| 48 | 26 | 1628 | 2496 | 2980 | 4x(407, 624, 745) |
| 48 | 27 | 1575 | 2592 | 3033 | 9x(175, 288, 337) |
| 48 | 28 | 1520 | 2688 | 3088 | 16x(95, 168, 193) |
| 48 | 29 | 1463 | 2784 | 3145 | Primitive |
| 48 | 30 | 1404 | 2880 | 3204 | 36x(39, 80, 89) |
| 48 | 31 | 1343 | 2976 | 3265 | Primitive |
| 48 | 32 | 1280 | 3072 | 3328 | 256x(5, 12, 13) |
| 48 | 33 | 1215 | 3168 | 3393 | 9x(135, 352, 377) |
| 48 | 34 | 1148 | 3264 | 3460 | 4x(287, 816, 865) |
| 48 | 35 | 1079 | 3360 | 3529 | Primitive |
| 48 | 36 | 1008 | 3456 | 3600 | 144x(7, 24, 25) |
| 48 | 37 | 935 | 3552 | 3673 | Primitive |
| 48 | 38 | 860 | 3648 | 3748 | 4x(215, 912, 937) |
| 48 | 39 | 783 | 3744 | 3825 | 9x(87, 416. 425) |
| 48 | 40 | 704 | 3840 | 3904 | 64x(11, 60, 61) |
| 48 | 41 | 623 | 3936 | 3985 | Primitive |
| 48 | 42 | 540 | 4032 | 4068 | 36x(15. 112, 113) |
| 48 | 43 | 455 | 4128 | 4153 | Primitive |
| 48 | 44 | 368 | 4224 | 4240 | 16x(23, 264, 265) |
| 48 | 45 | 279 | 4320 | 4329 | 9x(31, 480, 481) |
| 48 | 46 | 188 | 4416 | 4420 | 4x(47. 1104. 1105) |
| 48 | 47 | 95 | 4512 | 4513 | Primitive |
| 49 | 1 | 2400 | 98 | 2402 | 2x(1200, 49, 1201) |
| 49 | 2 | 2397 | 196 | 2405 | Primitive |
| 49 | 3 | 2392 | 294 | 2410 | 2x(1196. 147. 1205) |
| 49 | 4 | 2385 | 392 | 2417 | Primitive |
| 49 | 5 | 2376 | 490 | 2426 | 2x(1188, 245, 1213) |
| 49 | 6 | 2365 | 588 | 2437 | Primitive |
| 49 | 7 | 2352 | 686 | 2450 | 98x(24. 7. 25) |
| 49 | 8 | 2337 | 784 | 2465 | Primitive |
| 49 | 9 | 2320 | 882 | 2482 | 2x(1160, 441, 1241) |
| 49 | 10 | 2301 | 980 | 2501 | Primitive |
| 49 | 11 | 2280 | 1078 | 2522 | 2x(1140, 539, 1261) |
| 49 | 12 | 2257 | 1176 | 2545 | Primitive |
| 49 | 13 | 2232 | 1274 | 2570 | 2x(1116. 637. 1285) |
| 49 | 14 | 2205 | 1372 | 2597 | Primitive |
| 49 | 15 | 2176 | 1470 | 2626 | 2x(1088. 735. 1313) |
| 49 | 16 | 2145 | 1568 | 2657 | Primitive |
| 49 | 17 | 2112 | 1666 | 2690 | 2x(1056, 833, 1345) |
| 49 | 18 | 2077 | 1764 | 2725 | Primitive |
| 49 | 19 | 2040 | 1862 | 2762 | 2x(1020. 931. 1381) |
| 49 | 20 | 2001 | 1960 | 2801 | Primitive |
| 49 | 21 | 1960 | 2058 | 2842 | 98x(20. 21, 29) |
| 49 | 22 | 1917 | 2156 | 2885 | Primitive |
| 49 | 23 | 1872 | 2254 | 2930 | 2x(936, 1127, 1465) |
| 49 | 24 | 1825 | 2352 | 2977 | Primitive |

| $m$ | $n$ | $a = m^2 - n^2$ | $b = 2mn$ | $c = m^2 + n^2$ | *Primitive or multiple* |
|---|---|---|---|---|---|
| 49 | 25 | 1776 | 2450 | 3026 | 2x(888, 1225, 1513) |
| 49 | 26 | 1725 | 2548 | 3077 | Primitive |
| 49 | 27 | 1672 | 2646 | 3130 | 2x(836, 1323, 1565) |
| 49 | 28 | 1617 | 2744 | 3185 | Primitive |
| 49 | 29 | 1560 | 2842 | 3242 | 2x(780, 1421, 1621) |
| 49 | 30 | 1501 | 2940 | 3301 | Primitive |
| 49 | 31 | 1440 | 3038 | 3362 | 2x(720, 1519, 1681) |
| 49 | 32 | 1377 | 3136 | 3425 | Primitive |
| 49 | 33 | 1312 | 3234 | 3490 | 2x(656, 1617, 1745) |
| 49 | 34 | 1245 | 3332 | 3557 | Primitive |
| 49 | 35 | 1176 | 3430 | 3626 | 98x(12, 35, 37) |
| 49 | 36 | 1105 | 3528 | 3697 | Primitive |
| 49 | 37 | 1032 | 3626 | 3770 | 2x(516, 1813, 1885) |
| 49 | 38 | 957 | 3724 | 3845 | Primitive |
| 49 | 39 | 880 | 3822 | 3922 | 2x(440, 1911, 1961) |
| 49 | 40 | 801 | 3920 | 4001 | Primitive |
| 49 | 41 | 720 | 4018 | 4082 | 2x(360, 2009, 2041) |
| 49 | 42 | 637 | 4116 | 4165 | 49x(13, 84, 85) |
| 49 | 43 | 552 | 4214 | 4250 | 2x(276, 2107, 2125) |
| 49 | 44 | 465 | 4312 | 4337 | Primitive |
| 49 | 45 | 376 | 4410 | 4426 | 2x(188, 2205, 2213) |
| 49 | 46 | 285 | 4508 | 4517 | Primitive |
| 49 | 47 | 192 | 4606 | 4610 | 2x(96, 2303, 2305) |
| 49 | 48 | 97 | 4704 | 4705 | Primitive |
| 50 | 1 | 2499 | 100 | 2501 | Primitive |
| 50 | 2 | 2496 | 200 | 2504 | 8x(312, 25, 313) |
| 50 | 3 | 2491 | 300 | 2509 | Primitive |
| 50 | 4 | 2484 | 400 | 2516 | 4x(621, 100, 629) |
| 50 | 5 | 2475 | 500 | 2525 | 25x(99, 20, 101) |
| 50 | 6 | 2464 | 600 | 2536 | 8x(308, 75, 317) |
| 50 | 7 | 2451 | 700 | 2549 | Primitive |
| 50 | 8 | 2436 | 800 | 2564 | 4x(609, 200, 641) |
| 50 | 9 | 2419 | 900 | 2581 | Primitive |
| 50 | 10 | 2400 | 1000 | 2600 | 200x(12, 5, 13) |
| 50 | 11 | 2379 | 1100 | 2621 | Primitive |
| 50 | 12 | 2356 | 1200 | 2644 | 4x(589, 300, 661) |
| 50 | 13 | 2331 | 1300 | 2669 | Primitive |
| 50 | 14 | 2304 | 1400 | 2696 | 8x(288, 175, 337) |
| 50 | 15 | 2275 | 1500 | 2725 | 25x(91, 60, 109) |
| 50 | 16 | 2244 | 1600 | 2756 | 4x(561, 400, 689) |
| 50 | 17 | 2211 | 1700 | 2789 | Primitive |
| 50 | 18 | 2176 | 1800 | 2824 | 8x(272, 225, 353) |
| 50 | 19 | 2139 | 1900 | 2861 | Primitive |
| 50 | 20 | 2100 | 2000 | 2900 | 100x(21, 20, 29) |
| 50 | 21 | 2059 | 2100 | 2941 | Primitive |
| 50 | 22 | 2016 | 2200 | 2984 | 8x(252, 275, 373) |
| 50 | 23 | 1971 | 2300 | 3029 | Primitive |
| 50 | 24 | 1924 | 2400 | 3076 | 4x(481, 600, 769) |
| 50 | 25 | 1875 | 2500 | 3125 | 625x(3, 4, 5) |
| 50 | 26 | 1824 | 2600 | 3176 | 8x(228, 325, 397) |
| 50 | 27 | 1771 | 2700 | 3229 | Primitive |
| 50 | 28 | 1716 | 2800 | 3284 | 4x(429, 700, 821) |
| 50 | 29 | 1659 | 2900 | 3341 | Primitive |
| 50 | 30 | 1600 | 3000 | 3400 | 200x(8, 15, 17) |
| 50 | 31 | 1539 | 3100 | 3461 | Primitive |
| 50 | 32 | 1476 | 3200 | 3524 | 4x(369, 800, 881) |

| $m$ | $n$ | $a = m^2 - n^2$ | $b = 2mn$ | $c = m^2 + n^2$ | *Primitive or multiple* |
|---|---|---|---|---|---|
| 50 | 33 | 1411 | 3300 | 3589 | Primitive |
| 50 | 34 | 1344 | 3400 | 3656 | 8x(168, 425, 457) |
| 50 | 35 | 1275 | 3500 | 3725 | 25x(51, 140, 149) |
| 50 | 36 | 1204 | 3600 | 3796 | 4x(301, 900, 949) |
| 50 | 37 | 1131 | 3700 | 3869 | Primitive |
| 50 | 38 | 1056 | 3800 | 3944 | 8x(132, 475, 493) |
| 50 | 39 | 979 | 3900 | 4021 | Primitive |
| 50 | 40 | 900 | 4000 | 4100 | 100x(9, 40, 41) |
| 50 | 41 | 819 | 4100 | 4181 | Primitive |
| 50 | 42 | 736 | 4200 | 4264 | 8x(92, 525, 533) |
| 50 | 43 | 651 | 4300 | 4349 | Primitive |
| 50 | 44 | 564 | 4400 | 4436 | 4x(141, 1100, 1109) |
| 50 | 45 | 475 | 4500 | 4525 | 25x(19, 180, 181) |
| 50 | 46 | 384 | 4600 | 4616 | 8x(45, 575, 577) |
| 50 | 47 | 291 | 4700 | 4709 | Primitive |
| 50 | 48 | 196 | 4800 | 4804 | 4x(49, 1200, 1201) |
| 50 | 49 | 99 | 4900 | 4901 | Primitive |

# Further References

Barker, Andrew, ed. *Greek Musical Writings*. Volume 2: *Harmonic and Acoustic Theory*. Cambridge: Cambridge University Press, 1989.

Boethius, Anicius Manlius Severinus. *Fundamentals of Music*. Translated by Calvin M. Bower. New Haven, CT: Yale University Press, 1989.

Cornford, Francis MacDonald. *Plato's Cosmology: The Timaeus of Plato Translated with a Running Commentary*. London: Routledge, 1937.

Eves, Howard. *Great Moments in Mathematics (Before 1650)*. Washington, DC: Mathematical Association of America, 1980.

Hall, H. S., and F. H. Stevens. *A School Geometry*. New York: Macmillan, 1960.

Heath, Thomas L., trans. *Euclid's Elements*. Santa Fe, NM: Green Lion Press, 2003.

Glenn, William H., and Donovan A. Johnson. *The Pythagorean Theorem*. New York: McGraw-Hill, 1960.

Guthrie, W. K. C. *A History of Greek Philosophy*, Volume 1: *The Earlier Presocratics and the Pythagoreans*. Cambridge: Cambridge University Press, 1962.

Loomis, Elisha Scott. *The Pythagorean Proposition*. Reston, VA: National Council of Teachers of Mathematics, 1968.

Maor, Eli. *The Pythagorean Theorem: A 4,000-Year History*. Princeton, NJ: Princeton University Press, 2007.

Nelson, Roger B. *Proofs without Words*. Washington, DC: Mathematical Association of America, 1993.

———. *Proofs without Words II*. Washington, DC: Mathematical Association of America, 2000.

Posamentier, Alfred S. *Advanced Euclidean Geometry.* Hoboken, NJ: John Wiley, 2002.

Posamentier, Alfred S., Robert L. Bannister, and J. Houston Banks. *Geometry: Its Elements and Structure.* New York: McGraw-Hill, 1977.

Posamentier, Alfred S., and Ingmar Lehman. *The Fabulous Fibonacci Numbers.* Amherst, NY: Prometheus Books, 2007.

———. *π: A Biography of the World's Most Mysterious Number.* Amherst, NY: Prometheus Books, 2004.

Posamentier, Alfred S., and Charles T. Salkind. *Challenging Problems in Geometry.* New York: Dover, 1988.

Sierpinski, Waclaw. *Pythagorean Triangles.* New York: Yeshiva University Press, 1962.

Swetz, Frank J., and T. I. Kao. *Was Pythagoras Chinese?* University Park: Pennsylvania State University Press, 1977.

# Index